James R. Downey

WINNING THE WAR

Also by COLONEL JOHN B. ALEXANDER

Future War

COLONEL JOHN B. ALEXANDER, U.S. ARMY [RET.]

WINNING THE WAR

Advanced Weapons,
Strategies,
and Concepts
for the Post-9/11 World

THOMAS DUNNE BOOKS
≈ ST. MARTIN'S PRESS
NEW YORK

THOMAS DUNNE BOOKS.
An imprint of St. Martin's Press.

Library of Congress Cataloging-in-Publication Data

Alexander, John B.
    Winning the war : advanced weapons, strategies, and concepts for the post-9/11 world / John B. Alexander.—1st ed.
        p.   cm.
    ISBN 0-312-30675-X
    1. United States—Armed Forces—Weapons systems.   2. Military art and science—United States—History—21st century.   3. War on Terrorism, 2001–.   4. Nonlethal weapons—United States.   5. Military doctrine—United States.   6. United States—Military policy.   7. World politics—21st century.   I. Title.

UF503.A42   2003
355.02'0973'0905—dc21

                                                                            2003041380

First Edition: August 2003

10   9   8   7   6   5   4   3   2   1

Dedicated to *The Quiet Professionals*
and the warriors who understand
firsthand the meaning:
"For those who have fought for it,
life has a special flavor the
protected will never know."

CONTENTS

ACKNOWLEDGMENTS

As with any work of this complexity, I have many people to thank. They have contributed in many ways. Some provided information, while others read drafts and made improvements through their thoughtful comments.

My appreciation for these efforts is presented in no particular order. Thanks go to Tom Clancy, General Carl Stiner, General Pete Schoomaker, Burt Rutan, Col. Hawk Holloway, Drs. Kit and Kristin Green, Captain Sid Heal and Sergeant Bob Alcaraz (Los Angeles sheriff's department), Rick and Tom Smith, Ed Vasel, Col. George Fenton, Col. Andy Mazzara, Lt. Doug Taylor (75th Rangers), Superintendent Colin Burrows, Jim Wes, Lt. Gen. Gordon Sumner, Lt. Gen. Dick Trefry, CSM Eric Haney, Dr. Johndale Solem, Dr. Gregg Canavan, Dr. Marty Piltch, Dr. Mike Caluda, Robert Bigelow, John Northrup, Lt. Jay Kehoe, Dr. Klaus-Dieter Thiel, Harry Rosen, Kathleen Hollingsworth, Dr. Fadi Essmael, Carlos and Frederico Aguiar and the Condor staff, Bud Kinsey, Dr. Art Schneider, Joe Fuller, Skip Lunsford, Dr. Bill Isbell, all members of Raytheon/General Dynamics Team Focus Vision of the Future Combat System competition, Captain Sher Bahadur Pun (Second Bn., Gurkha Rifles), Sameh Taha, Joe Hylan and the NDIA organization that has sponsored the Non-Lethal Defense conferences of the past decade, and my son, Sergeant Marc Alexander (Palm Beach sheriff's department). Dr. Miriam John and the entire panel of the National Research Council Study on Non-Lethal Weapons and Technology deserve my thanks and yours for their efforts to get the NLW program on track for the long-term.

Special appreciation goes to Dr. Dean Judd, Dr. Hal Puthoff, and Dr. Eric Davis, each of whom drafted a segment on alternative energy sources. Any errors in that material are my responsibility.

Recognizing much of what is written herein will be highly controversial, I emphasize that my recognition of these people and organizations does not necessarily constitute their agreement with what I have stated. In some cases, we agree to disagree but do so with academic scholarship.

Again, I wish to express my gratitude to my editor at St. Martin's Press, Pete Wolverton, his assistant, John Parsley, and my agent, Ralph Blum. Finally, I want to thank especially my beautiful wife, Victoria, for her contributions in being the first to read each draft and catching many of my grammatical errors.

RAID

"ALPHA DELTA, THIS IS *Sierra One-One. Target confirmed,"* Sergeant First
Class Bentley stated calmly from his concealed position 800 meters from the road
leading northeast out of Tuzla. He was observing a small convoy winding down
a two-lane highway. It consisted of several old SUVs—a favorite vehicle around
the world—and two dilapidated Yugos. There were also two Soviet-era BMP
light armored vehicles. One BMP was leading the convoy and the other trailing.
Exactly in the middle was a brown school bus with the windows painted black.

Through his binoculars SFC Bentley could see several people sitting on top of
the lead BMP. Careful study revealed that one of them was an American soldier
with his arms tied behind him. Unlike the others riding on the BMP, he wore
no flak jacket or helmet. Appearances could be misleading. While SFC Bentley
had photos of all of the kidnapped soldiers, positive identification could not be
made at that range. "Probable friendlies," he reported over the radio. That was
good news for Brig. Gen. Tom Swift, the special operations commander who was
running the operation from Stuttgart, Germany. When you were operating
against isolated guerrilla forces, it was unlikely that the adversary had the sophisticated equipment necessary to intercept or decrypt the secure radio communications between U.S. elements.

Five days earlier, a specially trained terrorist unit from Serbia had infiltrated
into Bosnia looking for Americans to kidnap. After careful planning, they captured a squad of soldiers who had grown lax in their security procedures while
on routine patrol. Assigned to Task Force Eagle, they were members of the 1st
Battalion, 155th Infantry, from the Mississippi National Guard who had again
volunteered for the multinational Stabilization Force (SFOR) in Bosnia.

Though there had been numerous warnings about attempts to kidnap Americans, few ever occurred. Those had been individuals who had slipped off the
base to drink and were lured into traps by young females offering what testos-
terone dictates. Never before had a unit on patrol been overtaken and quickly

disarmed. The attack had gone down so fast that the squad barely had time to alert the command post that anything was wrong. This led to delays in establishing the fact that a kidnapping had taken place. The confusion provided the terrorists time to conceal their trail.

To make matters worse, two men and a young woman were found with their throats sliced. Unbeknownst to most of the local population, these three had been providing information to the military intelligence units responsible for this sector. The counterintelligence nets in the area had been penetrated.

Their having accomplished this recent mission meant the adversaries were highly trained, possibly by former Spetsnaz who were sympathetic to their cause. It also indicated there was still substantial local approbation by ethnic Serbs for the disgraced and imprisoned former Serbian leader, Slobodan Milosevic. The objective of the kidnapping was made clear almost immediately. They wanted to trade prisoners—the American soldiers for Milosevic and a few others.

Politically, the situation presented a problem on the home front. The American public had long since grown weary of peacekeeping operations. The brutality of ethnic cleansing in the Balkans had slipped from the limited attention span of all but a few. While there was a policy of not acceding to the demands of kidnappers, the thought of tortured soldiers returned piecemeal was unacceptable—especially with another election drawing nigh. Intelligence sources had revealed that the guerrillas had temporarily dispersed the prisoners. A direct attack would result in unacceptable civilian casualties. Even though many locals were providing support to the terrorists, the CNN factor would be too much to bear.

The break came when a phone call was intercepted indicating the timing of the convoy to bring the victims into Serbia proper. The prisoners would be all together and away from the congested city—ideal for the rescue operation. Feeling relatively secure, the terrorists were going to make the trip in daylight. Additional traffic on the roads and the notion that locals would warn them in the event of an ambush made this a viable option.

Inconspicuous Special Forces teams had been inserted into the area even before the kidnapping had occurred. Recruited for the mission for their native language speaking capability, they had been visiting bars, observing markets, and even doing odd jobs in the area. All the while, they were preparing for possible action. Placing SFC Bentley and his four-man team in their well-camouflaged temporary stone defensive position had gone without a hitch. In local conversations they had heard innuendos about Americans held prisoner but not enough to launch a rescue mission. Now their job would be to spot the convoy movement and alert both SFC Bentley and the Joint Special Operations Center (JSOC) forward command post as soon as possible. Given the demise of the local agents, these operators had to take additional precautions to ensure their survival. While they never made direct contact with one another, the fact was that they were constantly watching the surroundings and covertly signaling potential danger.

The operation was now set. One task would be to confirm that the convoy contained the Americans. Concern had been expressed that the terrorists might send out a decoy in daylight while making the real run at night. They could not risk a snatch operation and come up short. Though decades had passed, the experience of the Son Tay raid deep in North Vietnam always reminded the military of the need for current intelligence.

"This is Sierra One-One. Confirm location, Three-Two," reported SFC Bentley. That provided Brigadier General Swift with the exact location of the lead vehicle. He turned to the operations officer and gave the order to execute the first phase. Within a minute, the two F-18s that had been loitering several miles away dropped to the deck and came screaming over the city of Tuzla and directly toward the rear convoy. Coming fast and hot, they were not spotted by the terrorists until almost on top of them. Then, pulling up sharply and hitting the afterburners, the F-18s sent resounding shock waves through the distracted convoy.

The terrorists had anticipated being spotted but believed the hostages would provide protection from any overt attempts to bomb them. They were right but missed the real action. During the fast-mover fly-by a small micro–Unmanned Aerial Vehicle (UAV) had taken off from a concealed position near the roadbed. Nearly silent and less than six inches in diameter, it contained a two-ounce camera that could send back pictures through communications relays to the JSOC command post. Automated facial recognition technology had come a long way. The features of every service member were contained in a digital database along with DNA samples for casualty identification. Even though PFC Clement's face was badly swollen from several harsh beatings, there were sufficient details and the match was confirmed. With positive identification the rescue mission would go forward. "Execute phase two," said Brigadier General Swift, "and get cameras on station now." He was referring to small UAVs that would provide him with real-time multispectral images of the actions. For better or worse, small-unit action could be sent from the battle to the Pentagon or even the White House. Yes, *Brigadier General Swift thought,* the big CP in the sky and the ability to micromanage have gotten worse since the days of stacked helicopters in Vietnam.

Nearly simultaneously an AC-130 gunship, known as Spectre, appeared to the east of the convoy while another aircraft breached the hilltop about ten kilometers away. From news broadcasts of Operation Enduring Freedom the terrorists knew about the improved accuracy the Spectre brought to the battlefield. They were good but not good enough to snipe at vehicles with hostages present. However, the mission of the AC-130 was to be seen and to assist in sealing off the area should any reinforcements attempt to intercede in the rescue mission.

Distracted by the AC-130, the terrorists did not see the lower-flying CH-47 Chinook helicopter armed with the new advanced tactical laser that had been developed by Boeing. Firing photons, the laser weapon was accurate enough to destroy soft targets and yet intentionally miss humans in close proximity—even at extended distances. Within thirty seconds the laser beam danced through the convoy selectively puncturing tires and burning off antennas. In those few seconds the convoy was nearly at a standstill and only the BMPs maintained mobility.

Already in flight and loitering just over a nearby hill was another unmanned aerial vehicle—one especially designed for non-lethal weapons use. This version contained a payload of small gas-dispersing bomblets that could quickly cover the entire length of the convoy. In recent years calmative agents had improved dramatically. Given that both the terrorists and hostages represented a nearly homogeneous population in age and health factors, these chemical agents were sufficiently safe to employ on this operation. Certainly the extremely low probability of accidental death to the American hostages was well within operational acceptability. The terrorists didn't matter anyway.

The terrorists watched as the unexpected bomblets spewed out their numbing cargo. Within seconds everyone in the area was losing consciousness. They would be aroused shortly to a different world from that which they had left. Roles would be reversed.

The terrorist directly behind PFC Clement realized that something strange was happening. In response he began to raise his gun, intending to shoot the hostage. Before he could pull the trigger the top of his head departed, the result of a single sniper shot from Sergeant Phillips, who had been tracking events from the lair with SFC Bentley. With the advanced sniper rifle complemented by a laser target acquisition system that accurately measured minute changes in microclimatic conditions, 800 meters was a piece of cake. Actually, the use of the HRS .50-caliber weapon was a bit much. It could defeat light armor at 1,500 meters and had been brought along in case they had to shoot through some of the vehicles.

Within a minute the distinctive sound of low-flying Black Hawks could be heard as they, too, raced toward the ambush. With Comanche helicopter gunships in close support, the Second Platoon, Bravo Company, 2d Battalion, 75th Rangers fast-roped onto the target. The rotor wash had the added benefit of dissipating any lingering calmative agent so as to not impact the Rangers. The first objective was to snatch the Americans and evacuate them to Camp Eagle. Weapons at the ready, there was no resistance from the mostly unconscious band. PFC Clement was cut free and the other squad members were found tied and blindfolded in the bus. In less than three minutes the hostages were hoisted onto the two Black Hawks that had ever so briefly landed on the road and were now transporting the groggy ex-hostages to safety.

Next the Rangers turned their undivided attention to the terrorists. Three were identified from the worldwide database now available to law enforcement officers and counterterrorist military units. Adroitly the terrorists' hands were lashed behind them with simple but effective plastic cuffs. Several others were similarly restrained and dropped alongside the road. Rangers opened the fuel caps to each of the vehicles in the convoy and dropped in small time-delayed grenades.

Discombobulated, a few of the terrorists began to regain consciousness. Instinctively they reached for their weapons. Such moves were anticipated and these terrorists were quickly dispatched. A brief message from a Black Hawk pilot alerted the platoon leader to the crowd that was beginning to form to the west. They were attracted by the battle scene but were sympathetic to the terrorists. It was time to leave. "Sierra One-One, this is Blue Six. We're out of here; block the road." With that the concealed Special Forces team opened up with the M240C machine gun. Even at the slow rate of fire, 750 rounds a minute, it made an impression. With an effective range of 1,100 meters, the intent was not to hit anyone, just to demonstrate accuracy and keep the growing mob away.

With the Comanches circling, six Black Hawks swooped in and landed as the Rangers dashed for extraction. With them the terrorists were unceremoniously dragged across the field and slung into the birds. Each felt a pinprick in his thigh and lapsed back into unconsciousness. There would be no unruly behavior from them during the flight. In fact, it was fortunate that their intelligence value made them worth keeping.

A single Black Hawk settled down long enough to pick up the concealed Special Forces team. As they lifted off, explosions could be heard along the convoy. SFC Bentley wondered how many of the remaining terrorists had become sufficiently conscious to get away from the vehicles. However, he didn't worry about it.

As the United States was a nation at risk, the gloves had come off years before. Asymmetric engagements had become the norm, not the exception. Actuation of epidemic terrorism forever changed the rules for both sides. This raid *was but one almost insignificant incident in an intractably complex mosaic that held the destiny of Earth in the balance.*

THIS FICTIONAL vignette describes one possible future battle. The advanced technologies discussed are either currently deployed or under development. While the military must prepare for large battles, such as war in the Middle East or Korea, there will be many, many brief skirmishes dominating their attention. These actions will see small units engaging elusive foes and yet influencing regional stability. Decisions made by relatively junior officers or noncommissioned officers will have national consequences. Read now about the technologies of future conflict and the complex situations that must be confronted by the valiant men and women of our fighting forces.

What they really fear is a weak America.
—Caspar Weinberger

"WE HAVE SOME PLANES." With startling clarity that simple phrase punctuated the dark shadows that had been gathering for decades. It ushered in the formal beginning of a new kind of war. Uttered from the bloodied cockpit of American Airlines Flight 11 on that fateful September morn—a day that would from that time forward be known simply as 9/11—these words signaled the onset of an asymmetric response to America's military power. But the terrorists would soon live to regret their actions. **Ultimatum**—was President George W. Bush's succinct and poignant answer for a nation infuriated.

When war came in Desert Storm, the deadly efficiency of American-led coalition forces astounded all military analysts around the world. Few doubted that Western technology would be triumphant, but the battle succeeded beyond their wildest imagination. In mere minutes over the night skies of Iraq American Army and Air Force pilots established air superiority. Within four hours they had air supremacy, meaning they could fly at will with total impunity. Given the devastating defeat of Iraqi forces, most people have forgotten that Saddam's air defense system was far from third-rate. Rather, as Iraq was a client state of the former Soviet Union its forces contained their state-of-the-art equipment. A few Iraqi pilots launched to challenge the fighter aircraft of the unified forces. Quickly they and their comrades learned an unforgiving lesson. You fly—you die.

The target list had been carefully developed under the insightful guidance of Air Force Colonel John Warden based on his five-ring model for defeating an enemy.[1] Next the command and control structure was systematically destroyed, followed by attacks against critical elements of the infrastructure.

1

In short order Baghdad could not communicate with the troops in the field and the supply lines were cut off. Instant Thunder, Colonel Warden's plan to conduct parallel attacks with precision guided weapons against the enemy's centers of gravity, was executed with paralyzing effectiveness. In the first twenty-four hours of bombing, the air campaign took out more key targets than in any year of saturation aerial attacks during World War II.

Six weeks later, the ground campaign began delivering equal punishment. In less than 100 hours the battle was over. Iraqi armor that had held the Iranian forces at bay for eight years was no match for the swift-moving Abrams M-1 tanks that unexpectedly appeared out of the vast unnavigable desert. Shooting on the move with deadly accuracy, the M-1s, supported by attack helicopters and A-10 Warthogs, shredded the prepared defenses sending the Iraqi Army fleeing for their lives.

Desert Storm had a predictable downside. Thanks to the modern news media, pictures of the carnage were instantly available all around the world. Those countries that had been equipped by the Soviets knew they could not compete against American technology. The message to potential adversaries was quite clear: You don't want to fight the United States and its allies in open battle and on our terms. Thus it is reasonable to assume that the attacks of 9/11 have a causal relationship with our technologically overwhelming success in Desert Storm.

In many ways the United States has been slow to learn the fundamental lessons pertaining to application of power. There was no doubt about our technological superiority in Vietnam—yet we lost the war when we lost the political will to continue. While Desert Storm was a stunning military victory, a decade later Saddam Hussein was still in power and thumbing his nose at everyone. Technology alone is not the answer.

Worse yet was the improper employment of the military during the Clinton years. Simultaneous with the military services' being crippled through excessive budget cuts, they were sent off on missions forty-six times during 1992 through 1999. That was far more frequently than the sixteen commitments during the entire Cold War. For the most part the interactions were ineffective, and sometimes they were counterproductive. Probably the single worst example was the withdrawal of U.S. forces from Somalia following an ill-fated raid by Army Rangers and Delta Force members designed to capture the uncontrolled warlord Mohammed Farah Aidid. Disaster came as the direct result of poor policy decisions by the White House and secretary of defense that failed to provide adequate support for the missions assigned. The story became well known through Mark Bowden's stunningly accurate book *Black Hawk Down*[2] and was further popularized through director Ridley Scott's graphic depiction of events in the movie of the same name. The troops performed admirably. Irresponsible civilian leadership failed them.

As a result of the decision to withdraw by then-President Clinton, tens, if not hundreds, of thousands of people would die because of reluctance to place U.S. troops in harm's way. With minimal force, genocide in Rwanda could have been curtailed, a decision Clinton would publicly apologize for years later in a trip to Africa. At the time, we sat by until the pictures of bodies floating down the rivers became unbearable. As the situation in the Balkans deteriorated, policy makers again wavered while ethnic cleansing went unchallenged. Eventually it was decided to enter into a marginally nugatory air campaign—one in which politicians of various nationalities, more concerned with correctness than military effectiveness, determined targets.

During the same period, known terrorist organizations were becoming bolder. The response of the United States to various attacks was to lob cruise missiles toward suspected encampments. The effects were to unnecessarily deplete our supply of critical munitions while convincing our adversaries that we were afraid to engage them on the ground. With constrained budgets, no new contracts were established to replenish missile stocks and our active-duty military forces were drawn down to an unacceptably low level.

Concomitantly, the intelligence community was both reduced in strength and politicized. After an academician was chosen as Director of Central Intelligence, reporting bad news was not viewed as career-enhancing and the old hands left in large numbers. In the aftermath of 9/11 it would be learned that an opportunity to eliminate known terrorist Osama bin Laden years earlier had been turned down by the State Department. We would also learn that the clues to prevent the attack were available. There was just no one to piece the puzzle together.

Despite misuse and abuse by the leadership of the Executive Branch, the military did maintain some sense of direction and improvements in technology. Advances were made in precision guidance and sensor systems while the intelligence community struggled to stay abreast in communications technology that made collection increasingly difficult.

Following 9/11 came President Bush's declaration of a *War on Terror* and the beginning of Operation Enduring Freedom.[3] The first major objective was to remove the Taliban from power in Afghanistan and to attack the al Qaeda terrorist network that operated safely from the mountain complexes near Pakistan. It took a month to establish the necessary political alliances for basing and overflight rights. That time was also needed to position the logistical support and aircraft required to initiate a substantial air campaign.

At the forefront of the action inside Afghanistan were various elements of the U.S. Special Operations Command (USSOCOM). Operating in small units, special operations forces (SOF) were trained to function behind enemy lines, develop and coordinate guerrilla forces, and conduct direct action strikes against designated targets. Worthy of note is the fact that during the massive reductions of conventional units in the 1990s USSOCOM remained relatively level in personnel and funding. However, like other units they had been constantly committed for the past decade, and it was not unusual for some element of the command to be active in more than 150 countries in any given year.

It was a combination of special operations forces and airpower that was responsible for the apparent quick victory in Afghanistan. It was fortunate that most of the opposing Afghan forces chose to switch rather than fight. Being on the winning side is an old tradition in that part of the world. Most analysts overlooked the bad news—that these Afghans would switch back should that be to their advantage. Further, each tribe came with separate agendas, and not all were commensurate with creation of a national government. The best news for Americans was that U.S. casualties were extremely low. Unfortunately, more were killed and wounded due to misguided friendly ordnance than by enemy action. The big fights forecast for strongholds never materialized; a harbinger of changes in protocols in conflicts yet to come.

The other good news for the United States was that we did not have to place large ground forces in the theater. The reality is that except for Marine Expeditionary units and a few light Army elements, we could not have introduced a significant force into Afghanistan—certainly not an armored capability such as that of Desert Storm. First, we don't have them. Second, and more significant, you can't get there from here. Strategic lift capability is the long pole in every deployment plan.

Military forces aside, the composition of the new adversaries has dictated new objectives—ones that focus efforts on soft infrastructure targets in areas populated with noncombatants. In initiating the *War on Terror,* President Bush said this would be *a new kind of war.* He was correct, but it remains to be determined how the parameters are defined and whether we have the intestinal fortitude to prosecute such a conflict. It will cause us to take actions for which we have criticized others when they were faced with similar situations. Definition of the conflict is a nontrivial issue and one that is still in question. The determination of this president is detailed in Bob Woodward's book *Bush at War*.[4] It remains to be determined if he and the American people will see it through.

In my view we have slipped into World War X. I call it WWX because no one knows with certainty how many global conflicts have occurred and there is academic debate about how the Cold War should be viewed. This new war will be fought along the lines of belief systems, not merely nation-states. Pres. George W. Bush's idea of *Axis of Evil* falls short both geographically and conceptually. As an attempt to classify easily identifiable adversaries, it does not adequately accommodate the complexity of terrorism. Rather, it calls to account known adversaries that can safely evoke hatred in many Americans. It also reflects a *politically correct* view that all religions can accommodate the coexistence of differing belief systems and still function in a world that is dependent on worldwide trade and economic interests.

Unfortunately, the world is not that simple. While 9/11 was what futurist John Peterson calls a punctuation in history, it is only one of many events that will embroil us in a global conflict. The apparent military success in Afghanistan may disperse al Qaeda leadership temporarily, but it does little to address the underlying causes that allowed this network to gestate and flourish. Obsession with being politically correct will lead to failure and the deaths of many innocent civilians. There is hope provided we act decisively. However, those actions will call for uncompromising use of force and will also take the lives of noncombatants. The existing laws of war will be challenged and found to be inadequate.

America will be forced to make some very tough choices. If this is done with sufficient courage and foresight national survival is possible and our allies will also benefit from our position of strength. Dalliance will create an inextricable morass—one from which it will be difficult to emerge with our existing quality of life and concomitant civil liberties.

It is already incisively clear that America will have to act unilaterally in many cases if our national interests are to be protected. Alliances are transient and the United States is nearly unique in having transformed a heterogeneous population into a unified nation. In most areas of the world true allegiance stops at family, community, or possibly tribal borders. Nations are transient useful myths created for economic expediency, delineated by cartographers and warlords, and perpetuated by those seeking to obtain and project power.

Now emerges a true paradox for defense. As we are disproportionate users of natural resources, particularly energy, there is a need to simultaneously develop expanded ties in foreign economies while protecting the vulnerable infrastructure required to support our way of life. Those opposed to globalization view the United States as a voracious consumer, one that exploits weaker developing countries to support unsustainable growth. Developing nations will compete for those same resources, thus exacerbating an inevitable clash of cultures. While some propose that humans will spontaneously evolve as spiritually motivated beings, the evidence against that likelihood is overwhelming. Therefore, we are faced with the probability that use of force will become essential but that a deterrent capability will allow us to retain sovereignty while balancing our responsibilities to assist in the evolution of the rest of the world.

This book will provide an understanding of evolving weapons systems and concepts to be employed across the spectrum of conflict and beyond. The thesis accounts for a totally integrated approach that considers the intricate interrelationships between national security and global interests in the environment, social advancement in health and education enmeshed in reciprocal economy. It will call for judiciousness in use of force when such is prudent. This will address the significant advances that have occurred in non-lethal weapons since the writing of my previous book, *Future War*. It will expound on the ability to deliver overwhelming force on demand, including the ability to place U.S. forces on the ground quickly and safely and then sustain them for the duration of the operation. At the same time, it will explore the protection necessary to defeat asymmetric attacks against our infrastructure. And finally, it will propose that we unilaterally place weapons in space that will guarantee national security when speed and assurance of destruction are needed. Given the politically correct aversion to space-based missile defense assets, these technologically viable proposals will alter the way you view space.

"The thing that people around the world are always talking about is a too strong America and all that. What they really fear is a weak America." Caspar Weinberger, during an interview with Larry King following 9/11, spoke those words.[5] They are correct. Now we address what to do about it.

PART I

THE TOOLS OF WAR

THUNDER ROLLED ACROSS THE Arabian Peninsula, and in the wake of Desert Storm lay strewn a scene of unparalleled carnage. Seemingly invincible, American Forces had defeated a relatively sophisticated and experienced adversary. The ease and swiftness with which Iraq fell astonished all observers. Countries that relied on matériel from the former Soviet Union now knew their defenses were no match for Western forces.

The expensive arms buildup of the Reagan years had paid off handsomely. First, incapable of matching the combination of economic power and advanced strategic weapons systems, the Soviet Union hand crumbled and imploded. Now the superiority of Western technology employed in conventional combat was proven unequivocally. Air defenses were neutralized in hours. The combat-tested Iraqi Air Force fell from the skies every time they attempted to engage coalition forces. Even survival dashes to Iran frequently ended tragically. On the ground Iraq's expansive armored forces were systematically decimated. The thoroughly disheartened troops surrendered to the advancing army so fast that handling the vast numbers was a considerable problem.

From the video game–like qualities of pictures recorded by combat cameras the press coined the phrase *Nintendo War*. Spectacular displays of remotely designated precision-guided rounds destroying their targets captivated audiences. The night sky over Baghdad blossomed with antiaircraft rounds fired in futility as stealth fighters flew with impunity and ravaged Saddam's command and control infrastructure.

While both the military and weapons developers rejoiced over the efficiency of their modern weapons, shortcomings, usually not reported on television, were noted. Laser-guided bombs occasionally lost lock and went astray due to interference from clouds or smoke rising from earlier strikes. Terrain-following cruise missiles failed to locate targets because bombs had

9

destroyed tall buildings that would have provided key markers. Satellites did not provide the continuous coverage desired by field commanders. Of most concern was the ratio of casualties accidentally inflicted from the air on ground forces of the United States and its allies. Clearly, there was work to be done.

The need for better non-lethal weapons was confirmed by troops involved in a myriad of peace support operations. The Joint Non-Lethal Weapons Directorate (JNLWD) had harvested the low-hanging fruit of simple systems that could be quickly provided. Now it was time to explore more advanced technology.

This part of the book will provide an overview of the full range of weapons entering the American arsenal. In keeping with the premise that all of these weapons concern *pragmatic application of force,* we will range from non-lethal to hyperlethal systems needed to vanquish the wide array of threats now proliferating.

CHAPTER ONE

"PHASERS ON STUN"

EVER SINCE DISCUSSION OF nonlethal weapons began, the analogy with the world-famous *Star Trek* weapons has been pandemic. Captain James T. Kirk's line "Phasers on Stun" has been repeated as both headlines and in the text of most articles about these systems.[1] To be sure, the Phaser, with its ability to temporarily incapacitate sentient beings, destroy materials it strikes, and kill only if necessary, would be an ideal weapon. Of course, it uncannily allows 100 percent recovery of the targets for which the *stun* setting is selected. The Phaser is small, light, and versatile and holds seemingly infinite energy. Unfortunately, it doesn't exist. But that does not mean that attempts are not being made to develop such a system. Even if some pretty advanced thinking is perfected, the handheld weapon is a long way off.

In the past five years non-lethal weapons have taken some big leaps. The initial push was rightfully to get weapons into the hands of troops that were deployed on peace support operations. Even in the military, few people understand the complexity of transitioning a new technology into a fielded weapon. Pressured by exigent circumstances, developers are urged to accelerate new technologies. Once this is done, almost inevitably some unexpected consequence occurs and the developers are blamed for lack of diligence. Remember Agent Orange?

THE ADVANCED TACTICAL LASER

Totally eclipsing the entire Department of Defense non-lethal weapons budget is Boeing's emerging Advanced Tactical Laser (ATL). This program seems to be growing by leaps and bounds and transitioned from tens of millions of dollars to hundreds of millions of dollars in a single year. We must hope the technology will be able to keep pace.

The theoretical concept is easy. A large laser is placed aboard an aircraft, creating a photon-emitting gunship, not unlike the AC-130H Spectre. Instead of the death-bringing 105mm howitzer supported by various forms of rapid-firing guns, the ATL shoots a devastating light beam.

Of course the basic concept has been around at least since science fiction was in its infancy. Edward Teller, in 1967 as a member of the Air Force Science Advisory Board, suggested a fleet of aerial battleships.[2] The problem has been making it work. A description of employing lasers as non-lethal weapons can be found in *Future War*.[3] The notion of using high-energy lasers has been explored for decades. It was believed that they would be a logical weapon for air defense purposes. With lasers being fired at the speed of light, no aircraft could evade them once they had been acquired.

There were early successes. By the 1970s military tests proved that a laser could cut off the wing of an aircraft in flight. Of course these tests were conducted against drones, as the glide path of a plane without wings is like that of a rock. Despite all the excitement, there were two overarching problems. One was the size of the laser. Since lasers are inherently inefficient, they tend to be very large in order to generate sufficient power to do damage to a target. The second problem was air. As the laser beam propagated through the atmosphere it had a heating effect on the air molecules. This was exacerbated whenever fog or dust was in the air, and obscuration is very common on battlefields. Thus the beam becomes scattered and therefore ineffective. To project a laser beam through the relatively dense air meant an ever-increasing demand for power. Near sea level, the amount of power required was raising almost exponentially to reach beyond three kilometers.

The tactical problem on the ground was very simple. Lasers could be defeated easily. As missiles could reach much farther than three kilometers, attacking aircraft would have the standoff distance necessary to safely shoot at and eliminate the laser weapon before it could be engaged.

However, the same problems did not exist at high altitudes, and the Air Force kept working on the technologies. For years they proposed the Airborne Laser Laboratory (ALL). The mission was to show that lasers were viable weapons for an airborne missile defense system. As early as 1981 they successfully tested airborne carbon dioxide lasers against a drone, and twenty-six months later they shot down five Sidewinder air-to-air missiles. Unfortunately, those operational tests did not lead to further development. Despite attempts by researchers to keep the program alive, it remained unfunded.

With the Strategic Defense Initiative a new approach was taken. The Airborne Laser (ABL) developers mounted a Chemical Oxygen Iodine Laser

(COIL) on a larger aircraft. They also installed a new optical system capable of projecting a beam hundreds of kilometers while compensating for any atmospheric disturbance. These tests proved that laser weapons had the technology necessary to acquire an ICBM in boost phase. The ABL could also maintain the tracking necessary to deposit sufficient energy to bore a hole through the missile and destroy the warhead. Still, the system required a large aircraft such as a modified Boeing 747 to hold the laser. Also encouraging was that engineers demonstrated they could boost the power output. In just five years a 400 percent increase was accomplished.[4]

And then came the Advanced Tactical Laser. The intent of this Boeing lead project is to put a 300-kilowatt COIL laser on smaller aircraft such as the V-22 Osprey or a CH-47 helicopter. The developers claim that they will be able to place a four-inch spot at up to twenty kilometers. The ATL would operate at a wavelength of 1.315 microns. The laser, its chemical fuel, and the laser beam director would be sized to fit on an aircraft platform. For the initial version of the ATL, the targets would be selected by a human operator who views the scene through a separate aperture co-aligned with the laser beam director. The operator would control laser pointing through use of some sort of manual designator. In more advanced versions of ATL, target selection could also be accomplished automatically using target-recognition and tracking software.

There are several issues to be resolved. It is believed that the ATL, using advanced adaptive optics, can beat the atmospheric turbulence problems that have prevented lasers from propagating near the surface. This means a deformable mirror will compensate for the distortion with 341 actuators that change at a rate of 1,000 per second.[5] It is this enhanced accuracy that relegates the ATL to the non-lethal weapons category. In fact, if dwell time is increased this system becomes very lethal and hard targets such as weapons can be destroyed. It is hoped that through beam steering the operators will be able to selectively engage pinpoint targets and avoid humans who may be located in the immediate proximity. As early as 1999 technical sanity checks to determine if the ATL could destroy targets concluded that it could hit and melt a target at fifteen to twenty kilometers in a few seconds.[6]

The next problem is whether the military will accept the danger of having the fuel for the COIL on an aircraft. The COIL fuel is comprised of a number of caustic chemicals that require careful storage and handling. It is a mixture of hydrogen peroxide and potassium hydroxide that combines with chlorine gas and water to produce the chemical reaction to power the laser.[7] In general, caustic chemicals are not approved to be carried on aircraft.

Finally, there is the payload. The ATL is designed to strike multiple targets. In fact, the Boeing diagram for concept of operations shows the laser engaging seventy-two different targets in forty seconds. These targets vary from soft tires (thirty-two in eight seconds) to rockets, and machine guns that require two seconds per object. However, that's all there is. Forty seconds is the entire payload carried by this aircraft. Once depleted, the ATL must be flown to a rearming point. The conceptual question is whether the military will be willing to pay a huge amount of money to develop a system with such limited applications. There may be special missions that justify the expense of this system, and in an unusual twist U.S. Special Operations Command has become the program sponsor.

Recently battlefield lasers took another gigantic step forward when it was demonstrated that they could destroy incoming artillery shells in flight. As these rounds travel in excess of 1,000 miles per hour, acquiring them is very difficult. Then it is necessary for the laser to depose sufficient energy to destroy the round. While the laser was developed by TRW in the United States, the work was done for Israel for the purpose of hitting Katyusha rockets, which are often fired from Lebanon.[8]

THE ACTIVE DENIAL SYSTEM

The most notorious of the new non-lethal weapons is a millimeter wave device that projects a beam that results in pain. Once it was officially unveiled as the Vehicle Mounted Active Denial System (VMADS) in late February of 2001, the press had a field day with the news. The *National Post* headline read: "U.S. Energy Beam Lightly Scorches Protestors."[9] *Inside the Navy*, a normally service-friendly publication described the VMADS as a ray gun and focused on possible adverse health effects.[10] And the Gannett News Service called it "Crowd Control Cookery."[11]

In response to the attack on the USS *Cole,* consideration was given to placing this system on board naval vessels. Since they would not be placed on vehicles, the decision was made to shorten the name to the Active Denial System (ADS). In reality, the ADS, developed under the able program leadership of Kirk Hackett at the Phillips Laboratory at Kirtland Air Force Base in New Mexico, is a major step forward for non-lethal weapons. While the effective range has not been revealed, the ADS can reach out hundreds of meters. The effect—nearly instantaneous pain felt by targeted individuals—is well understood empirically. There is more to be learned on the biological mechanism of such heating and how people in groups will respond once exposed to the RF beam.

Contrary to the wild claims of unknowledgeable news writers, the millimeter wave beam does not threaten to cook victims who fail to get out of the way. Operating at about ninety-five gigahertz, the very short wavelength only penetrates the skin about one-sixteenth of an inch. This is sufficient to attack the pain receptors near the surface of the skin but is believed to do little physical damage.

In fact, Phillips Laboratory has done extensive testing to ensure that the weapon would be safe to use on humans. The possibility of eye damage is frequently mentioned in news articles. Most of the authors' information came from so-called experts who had never seen the ADS or been involved in any of the testing but still felt a need to give their personal opinions. Recognizing the potential for concern long before the press raised the issue, animal tests with direct injection by the beam were conducted. The power levels used were well above any that humans would be exposed to and failed to induce any permanent damage.

In the interest of personal knowledge I had an opportunity to experience the effects of a demonstration model. Like other tests of non-lethal weapons I've participated in, I would put this in the *one time learning category*. The beam hit my hand, causing immediate and substantial pain. As soon as the beam was removed, the pain ceased. That is exactly what is needed for motivating people through pain compliance. The effects that need to be studied include the response of individuals and groups when exposed to a system such as ADS. Operators must ensure that people have adequate avenues of exit and the system is not employed to impose punishment.

While billed as a device that can fit on an HMWWV, known to the public as a Hummer, the test model at Kirtland takes up two large vans. Power and cooling take up much of the room, and there is a large steerable antenna that sits on top. Plans are to reduce the size dramatically, and some optimists even believe that someday there will be a handheld version.

The ADS meets the critical test. It is far better than a rock. The beam will move across a crowd and quickly convince them to disperse. Simple countermeasures are unlikely. Covering oneself so there is no exposed skin restricts both the ability to see and movement. It also clearly demonstrates intent to the peacekeeping forces and would permit escalation in the level of force authorized.

In my view the ADS could play a significant role in combat in urban areas. As discussed elsewhere in this book, the probability of fighting in cites is extremely high. In cities the distance between adversaries is very short, varying from tens to a few hundred meters. One of the biggest dangers to troops is an enemy sniper. To acquire a target, the shooter must look down his sights. Even though a good sniper chooses a position with minimal exposure, he must get a direct view of his target. The ADS can be used to sweep an area. The beam is invisible, leaving no indication where it is striking. The pain initiated by this weapon is sufficient that once exposed, the sniper will not be able to maintain eye contact with his target. Further, the ADS operator does not need to know where the sniper is located except for his general area. As most casualties occur while troops are crossing open areas, the ADS will prove to be an effective area suppression weapon.

THE PULSED ENERGY PROJECTILE

There is another class of high-energy laser systems designed for antipersonnel application. While these are a long ways off, Col. George Fenton, USMC (Ret.), director of the JNLWD, has literally introduced these as "Phasers on stun" during several briefings.[12] These are systems designed not to cause damage directly but to produce a kinetic shock through a laser-induced plasma. One such proposed system being developed by the Mission Research Corporation is the pulsed energy projectile (PEP). PEP would utilize a pulsed deuterium-fluoride (DF) laser designed to produce an ionized plasma at the target surface. In turn, the plasma would produce an ultrasonic pressure wave that would pass into the body, stimulating the cutaneous peripheral afferent nerves in the skin to produce pain. It would also act on the peripheral efferent nerves to induce temporary paralysis. In addition, PEP will produce shrapnelless flash/bang effects that can distract, flash-blind, and disorient humans.

The proposed PEP system would accomplish this at ranges of up to 500 meters. However, thus far laboratory tests have only been conducted at about the one-foot range. Those produced 270 Joules and acoustic (pressure) levels of 170 dB, which is sufficient to cause the desired effects. In addition, the PEP is being considered for use as blunt munitions. Test measurements show that the impact on the body would be very significant when compared with other systems. They have recorded peak pressure over 300 pounds per square inch. This is more than 200 psi more than the PepperBall rounds discussed later, in " 'Dust 'Em; Then Bust 'Em!' " If PepperBall hurts, the PEP would probably be incapacitating. However, unlike beanbag shotgun rounds, the

PEP impact would not penetrate the rib cage. The developers believe that they will be able to control the impact pressure and have rheostatic effects. [13]

Considerable work will be required before this system is ready for prime time. Current designs also have the PEP on a dedicated HMMWV. The range must be greatly extended and bio-effects cataloged. The military will have to determine if they are willing to accept a system that employs DF, a very dangerous substance, on the battlefield.

RIDE THE BUFFALO

"Ride the buffalo," a quote attributed to Hans Marrero, former Chief Instructor, Hand to Hand Combat, for the U.S. Marine Corps, is well known to the thousands of people who have experienced the effects of the M-26 TASER®. While some assailants have been able to shrug off the electrical jolt from previous TASERs, the new law enforcement model will put you down.

For an excellent example of the M-26 in action see *Collateral Damage,* starring the epitome of toughness, Arnold Schwarzenegger. In the movie, Schwarzenegger becomes appropriately irate with some terrorists who had killed his wife and son. While attempting to avenge their senseless deaths, police enter the room and zap Arnie with the TASER. As would happen in real life, he goes down hard. Serendipitously, we met with Rick Smith, CEO of TASER International, and his father, Phillips Smith, chairman of the board, immediately following a screening of the movie. My wife, Victoria, reviews movies and we saw *Collateral Damage* before it was released. The Smiths had sent several TASER stun guns to Warner Brothers for the movie but had no idea about their use. We noted that they could not buy better advertising. [14]

One of the best-known cases of TASER use was that of Rodney King, who was videotaped as he was being subdued and subsequently beaten by Los Angeles police. King, like some others who did not respond as expected to the TASER, was on drugs at the time of his arrest. As the early Taser models from various dealers were based on pain compliance, many people who were chemically altered and impervious to shock could continue to attack. A few determined individuals proved that they could also fight through pain and still be dangerous to police officers.

The introduction of the M-26 by TASER International has brought about a major leap in non-lethal or less-than-lethal weapons available to the police. First they extended the Air Taser range. The original models employed compressed gas to project a pair of darts up to fifteen feet. That provided an advantage over the handheld TASER stun guns that required direct contact between the gun and the attacker. With the wires the user could stay well beyond striking distance with hand or club.

The second change is even more important. The power has been increased significantly. This TASER is *not* a pain compliance weapon. Rather, it employs electro-muscular disruption technology. That means that when shocked, the muscles involuntarily respond and contract so that the attacker goes down immediately.

The name M-26 is derived from the twenty-six watts of power and 162 milliamps that are transmitted over the wires. This can be compared with seven watts and fifty-seven milliamps in the early versions. Weighing only eighteen ounces, the M-26 is powered by eight standard AA batteries that easily fit into the grip of the weapon. Most come equipped with a laser aiming device, and the strike of one of the darts will be within a few millimeters of the spot.

As one of those who has "ridden the buffalo," I concur with the statement of Capt. Charles "Sid" Heal, head of the Los Angeles County Sheriff's Department Special Enforcement Bureau. When asked if he would be willing to try a second shot with the M-26, he retorted, "Once is curiosity. Twice is stupid." In short, it hurts. Rick Smith shot me while I was visiting their facilities of TASER International in Scottsdale, Arizona. He demonstrated one of the attributes of the weapon by placing an electrode on each foot. This showed that even if an individual is struck on the extremities, the M-26 is effective. Since I was wearing leather boots and the barbs were on the outside, the penetration of the electrical charge was more than adequately proven.

In fairness to Rick, he explained several times what was to be expected and offered me the opportunity to change my mind. Following Sid's edict, I opted for curiosity. With the laser pointer on my chest, Rick hit the trigger. The sensation was as if I had been hit on the top of both feet with a sledgehammer. The leg muscles stiffened instantly and I fell backward. In this case there was someone behind me to break the fall. Not using the automatic five-second discharge provided when used by law enforcement, Rick cut the power in about a second. However, that was more than sufficient time for me to fully appreciate what the system can do.

From the electricity flowing through the boots there were small burn marks on the top of my feet. Once the power had been switched off, the pain stopped immediately. Physical recovery took several seconds. Still, I would not have been able to run for a period of minutes and the legs felt similar to the aftereffects of major cramping in the calves. In less than twenty minutes I was sufficiently recovered to drive my car quite safely.

My sixteen-year-old son, Josh, was with me at the time. Even after observing my experience, he decided that he would like to try it out. No couch potato, Josh is extremely physically fit and can withstand far more pain than the average person. Once jolted he stiffened and fell back into my arms. He is sure that there was no way that he would be able to fight through the muscle failure and pain. The relative safety of the M-26 is demonstrated as I allowed Josh to ride the buffalo. Had I felt there was any potential danger to him, I never would have permitted him to try it.

In the United States the M-26 is gaining in popularity with more and more police departments. Some departments use the TASERs as specialty weapons to be carried by supervisors or SWAT team members. In other departments TASERs are more widely distributed. Recently the Ohio State Police ordered enough M-26 TASERs to have one in every car.

When I served in the Dade County Sheriff's Department many years ago each officer carried a gun, handcuffs, and a few extra bullets. Today the police officer's strap-on gear is very extensive, including radios, pepper spray, a stick, and other items. In addition, body armor is the norm. Therefore, convincing officers to voluntarily add new equipment means that the item must have value to them. Those who have tried the M-26 TASER in an operational setting believe in the system enough to carry it.

The news media are poorly informed about the use of most non-lethal weapons and particularly the TASER. There have been several high-profile cases with headlines reading: "Man Dies After Police Use TASER," or words to that effect. While this is a true statement, people also die after reading newspapers, but that does not infer a causal relationship. There has never been a death attributed to a TASER in law enforcement. These reported deaths normally occur fifteen minutes or more after being shocked, never at the time of the incident. In all situations in which in-custody deaths have occurred following the use of a TASER, there was an underlying medical problem that brought about death. The single most common event is drug overdose. The overdose can be accidental, or in some cases people have swallowed balloons filled with narcotics. On occasion these balloons or condoms have broken, releasing large amounts of the chemical into the blood system of the victim. Death is not necessarily rapid when this happens. However, many of them display the symptoms of excited delirium or hyperthermia. Both induce exotic behavior. The subject is often irrational and violent. If hyperthermia occurs, the body is overheating and the person will take off his clothes, even in extremely cold weather.

Such erratic behavior frequently results in calls for law enforcement. When they arrive the suspect seems out of control. If lucky, the police are armed with a weapon such as a TASER, which can permit the suspect to be temporarily disabled so that he may be taken into custody and transported to medical assistance. If death follows it is not the result of the TASER. In fact, leading cardiologists have stated that there is no "pathophysiological mechanism whereby the application of TASER-like electrical stimulus anywhere on the body surface could be the cause of that person's death some forty-five minutes later."[15]

In preparing for a presentation to the International Association of Chiefs of Police (IACP) in Brasilia I was surprised to learn just how complicated non-lethal weapons politics has become. I was informed that non-lethal weapons were a fairly new topic in South America and that I should address what they were and how law enforcement agencies could use them. My first thought was to take some of my weapons along for demonstration purposes and I began to make arrangements. Since senior police officials organized the conference I thought there should be few problems.

Shortly I learned just how wrong that was. To take even a single system in I would have to become a licensed arms dealer. Further, I would need both import and export licenses. Tom Smith, president of TASER International, was kind enough to provide me with an education on the subject. It turned out that TASERs are banned weapons in Brazil and mine would have been confiscated upon arrival.

Congress, in its infinite wisdom, has enacted laws pertaining to the export of U.S. weapons. They are very sweeping and put non-lethal weapons in the same category as if we were exporting nuclear devices. When and if an export license is issued by the Department of Commerce, it specifies the dealer to receive the weapons and that they may not be resold to other dealers or exported to third countries and limits the quantity that can be made available.

While this is probably an unintended consequence of the act, such restrictions actually run counter to our beliefs and intent. Two undesirable effects occur as a result of barring U.S. non-lethal weapons from export. First, it limits the availability of life-conserving weapons to Third World countries. Many of them are faced with severe civil instability and need all the options they can get. By disallowing non-lethal weapons we force them to use their tried-and-true approach—shooting people.

In preparing me for the trip to Brazil my hosts sent a list of problems that they encountered quite frequently and for which non-lethal weapons might be useful. High up on the list were the well-publicized soccer riots that are so bad they frequently make international news. Other issues included prison riots, which normally end up with many people dead. The prisoners are normally on the losing end. Other politically sensitive issues included battles over agricultural land and the continuing conflicts with indigenous peoples in territorial disputes.

The second problem associated with the extreme export controls is that they put U.S. businesses at a disadvantage in international competition. This is not an issue unique to non-lethal weapons, as I have heard business owners in many sectors complain about the arcane bureaucracy established by the U.S. government. The effects do impact us all as they diminish revenues to the United States and contribute to our negative trade imbalance.

The basic resistance to export and use of electrical weapons systems has to do with human rights issues and allegations of torture. While it is true that misapplication of the TASER can happen, it is not the weapon that is inherently bad. Electricity as a means of torture has been around for a long time. For many years crank-driven telephones and radios were notorious for that purpose. The victim would be strapped down and electrodes placed on sensitive parts of his body while another person turned the hand generator. The pain was very severe and often yielded confessions. However, there was never a move to ban telephones.

Further, there has always been debate about the reliability of information derived from torture. Once induced, people are likely to say anything they think the interrogator wants to hear, whether it is true or not. TASERs in particular and non-lethal weapons in general have been at issue in this debate. Most surprising, it has been groups such as the American Civil Liberties Union, the International Red Cross, and many other peace-espousing organizations that have been the most vocal opponents of non-lethal weapons.

This was made very clear in the October 1999 issue of *The Futurist.* I provided an article covering all sides of the topic of non-lethal weapons. When the magazine was published I learned that Nick Lewer of the Bradford University Institute of Peace Studies had also been asked to write an article, but one supporting only the negative side of these systems. When these articles were published, the editor included a picture of a Tibetan monk holding a number of electrical shock devices that had been used on him while he was in a Chinese prison. In fairness to Nick, I found out at a

conference we were attending in Edinburgh, Scotland, that he had not provided the picture, even though it was supporting his article.

Since that time I have used his picture side by side with one that I took.
In my photograph, my wife, Victoria, is shown holding instruments of tor-
ture. Included are garden shears, a hammer, a drill, and a variety of other
household devices. My personal favorite is the most common torture tool.
In fact, I do support a ban on the export of these objects. However, the
American tobacco lobby will never let Congress ban the sale of cigarettes
abroad. The fact is that there is no object that cannot be used for torture.
It is the intent of the user, not the device, that is important.

There are steps that have been taken to mediate against misuse of some
stun guns. TASER International has long had a method for identifying the
specific weapon that had been used. Each dart cartridge contained a large
number of very small paper dots with individual identity codes. When fired,
the darts struck the suspect, but these dots scattered about the area.
Therefore, when police investigated they could call the company and determine the person who had been sold that cartridge.

Since then, TASER International has taken additional steps. In the base
of each M-26 TASER is a microchip. That microchip records every time
that the TASER is tested or fired. Many police departments keep logs on
every weapon. In that way law enforcement agencies can prove whether or
not their devices were used appropriately. This innovation provides additional security for both governments and citizens.

There are other applications for TASERs being developed. In recent years
there has been a furor over the use of land mines. Led most notably by the
late Princess Diana, Queen Noor of Jordan, and a host of celebrities, a hue
and cry for a ban has been heard round the world. However altruistically
intended, this very emotional response has fundamentally flawed logic. It is
true that there have been widespread problems with indiscriminate use of
small antipersonnel mines with no mechanism to incapacitate them and
scattered unexploded ordnance. However, the problem has been misidenti-
fied. The issue is improper use, not evil technology. Recent advances permit
mines to have limited life or to be controlled remotely. These advances can
be applied and we can still be empathic to the thousands of unfortunate
people who have been killed or maimed.

There are a number of missions for which mines are appropriate. As one
who has been chased on the ground, I find the idea of devices that can be
scattered and slow pursuit very appealing. There are many areas of the world
in which impenetrable barriers will become increasingly necessary. Since the

United States often will have small forces on the ground, mines offer effective measures for controlling terrain without spreading people too thin.

Non-lethal alternatives may fill some of the void. One development is a mine that can sense intruders and fire the TASER darts at them. Only temporarily incapacitating, the mine then sends a message so that the person can be captured. Other alternative systems are under consideration.

"DUST 'EM; THEN BUST 'EM!"

Pepper spray, or oleoresin capsicum (OC), has become a standard non-lethal weapon for law enforcement agencies. Today many officers carry spray dispensers on their belts and may have larger fogger devices in the trunk of their patrol cars. OC makes one hell of a party buster and will get most groups moving quickly for an exit.

While chemical agent sprays have earned a good reputation as low trauma–producing non-lethal weapons, two inherent problems have emerged with the use of pepper sprays. The small handheld spray cans meant the officer had to be in very close proximity to the suspect. Worse, sometimes OC doesn't work. There have been a number of suspects who barely blinked an eye even when the spray hit them directly. One California study indicates that as many as 15 percent of the population may not be affected by OC. The reasons for this are yet to be fully understood. A few brave souls have demonstrated they can actually eat the stuff right out of the canister.

One issue that may contribute to this statistic is the lack of industry standards pertaining to the amount and strength of the OC in the pepper spray. OC is commercially derived by processing oils from various species of hot peppers. Capsaicinoids are the actual ingredients within the pepper plant or OC that cause the burning sensation and inflammation of mucous membranes. From year to year the sunlight, rain, and growth nutrients can affect the "hotness" or capsaicinoid levels of the harvest of peppers that are used to make the OC. This is why each crop of peppers may have different levels of the active capsaicinoids such as capsaicin, homocapsaicin, dihydrocapsaicin, nordihydrocapsaicin, et cetera. And each capsaicinoid has a different ability to cause the burning or irritant sensation that makes Tabasco™ sauce and pepper spray so hot.

Therefore, products marketed as pepper spray can vary tremendously in effectiveness. Having lived in New Mexico, I became acquainted with the grading of chili peppers. As chili peppers are a food product, there are people whose job it is to test the chili peppers and determine just how hot they are to the taste. This taste test commonly involves grading by Scoville Heat Units or SHU rating. Because different tasters may judge a batch of OC oil differently, SHU can be a very subjective rating system. Most gringos are "smoked" by even a small amount of these peppers that are routinely used to flavor southwestern cuisine. However, taste buds will learn to accommodate hotter and hotter flavors, and there is indication that peppers may be mildly addictive.

From a manufacturer's perspective, it is far cheaper to use lower-grade OC rather than include higher concentrations of capsicum. Couple this with the fact that peppers can vary in hotness from crop to crop and it is easy to see that all pepper sprays are not equal. And some pepper spray brands may vary in hotness from year to year. Thus the competition for a wider market at lower costs has encouraged marketing of products with materials that are ineffective on many people. This variability in OC concentration and effect on target has led to a questionable reputation for OC in general.

PepperBall Systems are products of PepperBall Tactical Systems in San Diego. They have overcome both of the problems of a safer standoff distance and the variability in OC concentration and hotness. As a result, police departments and the military are increasingly using these systems in situations involving: domestic violence; public disturbance; suicide by cop; armed suspects; civil unrest; crowd control; jail riots; drug interdiction; barricaded suspects; hostage release; and high risk warrants. Innovatively, their patented PepperBall Technology has married concepts from the paint ball war game industry with chemical agent microencapsulation techniques to bring advanced chemical agent deployment products to the streets.

Their PepperBall System is a compressed gas launcher that shoots frangible plastic marble-size projectiles filled with powdered OC or VP (vanillyl pelargonamide) with amazing accuracy. Using a laser-aiming device, PepperBall launchers can hit an individual target at up to thirty feet. For area saturation coverage such as during a riot, the chemical agent projectiles can impact walls up to 140 feet away, thus providing increased safety for the officer using the launcher.

While most other non-lethal weapons have only a single effect (kinetic, chemical, electric shock, et cetera) on a suspect, the PepperBall projectiles are unique in that they produce a triple combination of effects. The PepperBall System produces: the psychological shock of being shot, the kinetic pain of the projectile impact, and the irritant effects of the chemical agent. Together this is referred to as the Chem-netics effect. Chemical effect

+ Kinetic effect = Chem-netics effect. This triple threat is extremely effective in gaining compliance from a suspect.

PepperBall projectiles travel at between 300 and 380 feet per second. The kinetic impact is substantial and will cause welts and bruising. The impact on the torso area fractures the projectiles, releasing the chemical irritant powder. Normally the officers fire several shots in rapid succession. This sends a small dust cloud of very strong irritant into the face of the suspect, who has been warned that a non-lethal weapon is about to be used on him. A normal response to the kinetic impact is to gasp for breath. This enhances the intake of the irritant powder into the eyes and lungs and is usually sufficient to gain compliance and allow the suspect to be disarmed or arrested.

Different launcher configurations are available. Since the non-lethal weapons use compressed gas, they are called launchers, not firearms. The SA200 semiautomatic launcher can hold up to 180 rounds in the hopper, which is useful when confronted by an angry crowd such as was the case in the Seattle WTO riots and at the Winter Olympics in Salt Lake City. The 3000-psi high-pressure air bottle ensures that the officer can keep shooting as long as necessary to engage multiple targets. In extreme situations, the SA200 has a riot pack configuration that allows a single officer to carry and shoot up to 850 projectiles without refilling compressed air!

For those close-quarters combat scenarios such as in a small room or for commercial aircraft and train security, the SA10 PepperBall launcher may be the best choice. It is a CO_2-powered ten-shot semiautomatic pistol that can launch a variety of non-lethal projectiles and can even be used to shoot a weapon out of a suspect's hand.

The plastic PepperBall projectiles are color-coded for easy identification. The primary projectiles are red-hot "tear jerkers." These are filled with either OC or VP powder in a proprietary formula developed for PepperBall Tactical Systems. In nature, both capsaicin I and VP (capsaicin II, nonivamide) are the hottest of the capsaicinoids. While OC powder contains many capsaicinoids, the VP formula contains only the single capsaicinoid, VP, as the active ingredient. This allows for extremely precise control of chemical agent irritant concentration and effect on target. Because the VP is pharmaceutically produced, it is very pure and very potent as a chemical irritant.

To ensure consistent chemical agent formula concentrations, HPLC (High Performance Liquid Chromatography) analysis is used to certify the capsaicinoid content of each product. This lab test is far superior to the very subjective SHU "taste test" used on other OC products.

The PepperBall technology goes a few steps further in offering a versatile system. PepperBall Tactical Systems has developed several rounds other than OC and VP for use in their launchers. One very useful round is the marking projectile. During riots, leaders are often spotted urging others on or organizing resistance to law enforcement efforts. By marking the key offenders with brightly colored paint, they can be photographed for identification as they move about the mob. This is useful to snatch teams, who can easily spot their suspect. While there are techniques for entering a crowd and apprehending selected leaders, they can be very risky to the officers. The marking agent lasts after the crowd has dissipated and the leaders can be arrested on warrants without the traumatic incidents attendant to removing them in front of their peers. Once isolated, most of the loudmouths lose their sense of bravado.

Two other rounds include water, to be used as a warning shot, and a glass-shattering round designed to break tempered glass in buildings or vehicles. There are occasions on which a suspect becomes stalled but remains locked in a car, refusing to comply with police requests. These glass-breaking rounds precede the use of the OC/VP rounds that are utilized to extricate the individual from the vehicle. This glass-breaking round has been used to reduce the typical freeway-clogging suspect standoff from hours to minutes.

A riot at the 2002 Winter Olympics showed that the kinetic impact alone from the water rounds was sufficient to gain compliance from an unruly crowd. Six officers with PepperBall riot systems moved hundreds of people several blocks with no injuries using water balls. They offer an advantage to the police, as they do not contaminate the suspects or the vehicle used to transport them to jail.

In 2002, PepperBall Tactical Systems advanced non-lethal options for law enforcement by introducing two new PepperBall Technology products: very potent VP pepper spray and powdered VP twelve-gauge projectiles that can be launched from most conventional shotguns. Law enforcement just got better.

DOES IT WORK?; OR, IT SEEMED LIKE A GOOD IDEA AT THE TIME!

After I signed a two-page release form in which I agreed that Ed Vasel, the inventor of the PepperBall technology and a vice president of PepperBall Tactical Systems, could do terrible things to my body, he and I stepped out behind the PepperBall building. In fairness to Vasel, he offered me several chances to opt out of the practical demonstration. Vasel also proposed that they provide me with a vest that would absorb some of the impact but still obtain the effects of the OC. Being not of sound mind, I decided to learn firsthand what a subject being apprehended would experience.

Bad choice! As the first of four rounds hit my chest, I instinctively started to spin away from the impact. Nevertheless, the first three rounds hit within an inch of one another. The instant onset of pain caused me to involuntarily take a deep breath from the cloud of OC that was engulfing my face. The combined effects had me on my knees holding my chest and coughing in less than two seconds. When I was asked questions about how I felt there was great difficulty in speaking for the next couple of minutes. However, within five minutes most of the effects had dissipated. Vasel has warned me that most people typically only welt or bruise. Depending on body fat content, some people may *leak* a little. I was a leaker. One projectile struck me directly on the bone near the sternum, causing a small abrasion. That bruise would take a while to totally disappear. The other welts cleared in a few days. While these effects did hurt, it was infinitely better than being shot by a Glock or Sig. The kinetic impact force that I experienced from the PepperBall system was only ten foot-pounds, as compared to a typical bean-bag extreme trauma impact of 120 foot-pounds. Chem-netics™ works!

IS OC SAFE?

There have been numerous complaints that OC on occasion can lead to death. As is the situation with TASERs, people have died after being exposed to OC. However, there have not been any cases in which the chemical agent was the cause of death. Noted years ago was that the practice of using a *hogtie* or *hobble* as a method of restraint did lead to deaths. Now banned in most police departments, the hogtie entailed using rope to tie a violent suspect's hands behind his back and then connecting the rope to his ankles. If placed facedown, some individuals experienced a compression of the chest leading to blocked airways and on occasion death. This became known as positional asphyxiation.

Because of these allegations, the National Institute of Justice undertook studies to explore the effects of OC when combined with other factors such as body weight, asthma or other pulmonary disease, use of respiratory inhaler medication, and even smoking.[16] The test was actually conducted by a team from the University of California, San Diego, School of Medicine and Medical Center. They found, as have all previous studies, that OC is safe to use on humans—even in combination with other restraints. The results also support the position of law enforcement agencies that have been subjected to lawsuits alleging "excessive force based on the unfounded contention that OC exposure results in respiratory compromise."[17]

While people sometimes feel as though they are not getting enough air, the oxygen intake is actually increased. This can lead to panic, but the sensation quickly subsides.[18] It also means that suspects focus on their bodily functions and are less inclined to continue aggressive behavior.

Children and guns has been an emotional issue for several years. In a *Washington Post* article I pointed out that non-lethal weapons, such as TAS-ERs and PepperBall, provide a viable alternative for home protection.[19] In an era in which many people feel a need to have weapons in their homes for self-protection, gun sales have increased. However, repeated studies have shown that children too frequently gain access to those weapons. Spurred on by violent television programs, even very young children are fascinated with guns and have proven quite ingenious at finding them. Of course gun manufacturers have urged that the weapons be securely locked up and trigger locks installed.

However, if one is sleeping in an area he feels is not secure, the necessity of retrieving locked weapons while an intruder is in the domicile is not acceptable. In such situations non-lethal weapons offer an alternative. Both the TASER and PepperBall products provide for a sufficient standoff to all allow a homeowner to engage a burglar and get away. In particular, a PepperBall launcher would offer multiple shots on multiple targets with the possibility of shots to the face should the intruder continue the attack. The objective is to *shoot and scoot,* not to apprehend the criminal. Leave that for the police. These non-lethal weapons can provide you and your family time to get out of harm's way.

Possibly of more importance is the safety these weapons provide in case of accident spawned by an inquisitive child. Should a youngster improperly use one of these non-lethal weapons, the worst that will happen is that he will feel sore for a short period of time. Despite some pain and discomfort, he will survive, but the same cannot be said for a firearm. These non-lethal weapons offer similar protection for adults. Too frequently we hear of a family member who mistakes another for an intruder and blows him away. Were a non-lethal weapon used for family protection, the question of intent to kill would be moot.

"REACH OUT AND TOUCH SOMEONE"

"You don't think rubber bullets are non-lethal, do you?"

As I passed through London's Heathrow Airport en route to Edinburgh, Scotland, that question was raised by the UK customs officer. After arriving on an international flight I was dashing for one on the domestic Midlands Airlines when he stopped me. While I was in a hurry, I noticed I was the only person clearing at that moment giving the officer more time for interrogation. My concern about making a relatively tight connection was not one of his. In fact, talking to me was about the only thing he had to do on this cloudy morning. He had first asked about my business in the United Kingdom and I told him I was attending two non-lethal weapons conferences. No, I was not an arms dealer, was my answer to his next question. Then he asked me to tell him what I considered to be a non-lethal weapon. Rather than a complex answer I responded with rubber bullets, thinking he would have an understanding of their use. Clearly, I was both right and wrong. His view of rubber bullets was colored by the news coverage from Northern Ireland. In fact, the IRA has waged a very successful psychological operations campaign based on fatalities and serious injuries caused by baton rounds.

The use of rubber bullets became popularized in the strife in Northern Ireland. There the use of these crude ballistically unstable projectiles by the British Army and Royal Ulster Constabulary (RUC) inevitably gave rise to deaths—between 1971 and 1989 a total of seventeen deaths were reported. This equated to one fatality for every 6,500 rounds fired. These tragic but inadvertent deaths received wide media attention, greatly influencing public perception about the use of these weapons.

The design of the baton rounds changed from PVC material to polyurethane went largely unreported. In 1994 a new weapon launcher was added to the inventory and the safety record improved substantially. Despite 13,500 baton rounds being fired since the introduction of the new weapon system, there have been no fatalities. Superintendent Colin Burrows of the RUC, commenting on the development and use of baton rounds, highlights the importance of not only technical design but also highly organized conflict management processes. Technology, combined with training and supervision, ensures that potentially life-threatening impacts may be substantially reduced.[20]

This is an important trend and points to a fundamental issue. Non-lethal weapons can improve and meet progressively higher standards of safety. The difficulty is in determining when initial deployment meets minimum standards knowing that while many lives may be saved with a new weapon system, a few may be lost as well. For law enforcement, legal woes from inevitable lawsuits following the few unintended deaths or injuries often dissuade agencies from trying alternative and innovative systems. Unfortunately, it is easier to explain deaths by firearms than even nonpermanent injuries from non-lethal weapons that have not been tested in the court system.

BETTER THAN A ROCK?

High technology is not the most important feature about non-lethal weapon—practicality is. Since the introduction of non-lethal weapons in the military, a basic question has been, "Is it better than a rock?" As simple as it sounds, the main concerns about these weapons are whether they are effective in accomplishing the mission and whether there is a simple countermeasure that people can use. Rocks are one of those countermeasures, and defeating them is a very important question.

Confronted with unruly crowds in peace support operations, soldiers have wanted weapons that permit them to stand back and apply force without resulting in killing people. As every person who has watched the news from the Middle East is aware, stones thrown by teenagers are very common and often lead to disastrous consequences. It is estimated that a person can throw a rock about 180 feet.

At present no direct-fire non-lethal weapons can engage a point target accurately at that distance. However, several sources are busily at work attempting to develop weapons that provide troops safe standoff. TASER International is experimenting with a round that can be fired without trailing wires behind it. There is a need for an officer to be able to engage more than one suspect in rapid succession. They are also developing a system that allows more than one round to be fired from the same weapon. Currently there is a need to replace the cartridge before a second set of darts can be deployed. This also cuts the power to the first target.

PepperBall Tactical Systems is developing a technique that will increase the effective range of the PepperBall projectiles. At issue is that increasing range usually means that you increase the velocity of the projectile. In so doing, there is greater impact on targets at closer ranges. Currently most PepperBall projectiles are safe at the muzzle. One negative effect from bean-bag rounds fired from standard shotguns is that they sometimes penetrate the rib cage, causing serious injury or death. Substantially increased velocity of the PepperBall projectiles could increase that risk. Therefore, PepperBall Tactical Systems is developing a system that can extend the range while keeping the impact within safe limits.

Other new versions soon will include a PepperBall launcher embedded in a high-power flashlight. An officer using his light will have the ability to use non-lethal force without the necessity of switching weapons. And for the consumers, there is a small device that looks like a key chain item. It carries one or two rounds powered by compressed gas. Concealed in your hand, it is ideal for protection while walking in a dark parking lot. The device is not gender-dependent and a variety of colors will be offered. Purple seems to be the most popular with women.

Whether either TASER or PepperBall technology gets to be better than a rock remains to be seen. What is known is that the capabilities of non-lethal weapons continue to evolve, continue to save lives, and continue to prevent wrongful death lawsuits. These weapons are not perfect, but they are getting better. We can shoot but not kill.

OTHER ADVANCES

There have been significant improvements in several other technologies that were first reported in *Future War*. While the Joint Non-Lethal Weapons Directorate harvested the low-hanging fruit quickly, other technologies remain in the research phase and some have found niche markets of their own. Note that considerable emphasis has been placed on stopping or deterring vehicles and watercraft. The following updates cover many of the changes.

LIGHTS

Among the most common non-lethal weapons are bright lights. Both the Department of Defense and law enforcement agencies need simple and effective systems, and lights serve multiple purposes. The recent changes are in improved illumination and control of the beam.

The Laser Dazzler™ system of LE Systems, Inc., is in limited production. It is a handheld battery-powered eye-safe, 532nm laser that is limited to 140 milliwatts, to ensure that it is eye-safe at the aperture. The same technology can be scaled up and applied to the protection of naval vessels, area denial, or a perimeter defense system.

The intent is to reduce risks of injury to both combatants and noncombatants by temporarily hindering the eyesight of the targeted personnel and providing visual high-brightness illumination to the area of interest. In development is a higher-power system with tunable output intensity to provide the necessary level of output light for non-lethal applications. The non-lethal laser light incapacitates the target by inflicting temporary loss of vision in the general direction of the illuminating laser. The current handheld Laser Dazzler has been designed to be eye-safe at the aperture, to all current FDA and ANSI standards, by limiting the output power to 2.55 milliwatts/cm².

Since an area defense platform will allow for a larger aperture system, LE Systems is working to incorporate an active range sensor that will prevent accidental viewing of the higher-power density beam at close range. The system will be lightweight, compact, and robust for mounting onboard a vehicle, a ship, or an airborne platform and will have an effective operating range of five to ten kilometers.

In addition, the five-watt Laser Dazzler will provide a high-intensity searchlight beacon. A highly concentrated source of green light can provide almost daylight conditions. A high-intensity laser, whose output can be highly collimated, will allow commanders to identify targets at longer ranges and take away the operator's ability to see while the threat is being analyzed. Since the 532nm wavelength has good transmission in water, the system can be used under fog and rain conditions.

The effort began on two fronts, first the design of a resonator that would be simple, compact, and efficient and second, selecting a laser diode package that would fit into a flashlight-size package and provide the 808nm wavelength required by the resonator design. The design of the resonator concentrated on developing various optical configurations that would maximize the conversion from the 808nm diode to 532nm. LE Systems worked with a number of diode suppliers to develop the 808nm package to deliver the beam in a form that interfaced efficiently with the LESI resonator. This effort resulted in a "deliverable laser resonator" that delivered in excess of 150 milliwatts, in a package that was 1.75 inches in diameter, and 8.50 inches long.

The second-generation design laser integrated the laser diode into the resonator and incorporated improvements in the coatings, crystals, and mechanical design. This design, which was incorporated into the first production prototype Laser Dazzlers, delivered in excess of 150 milliwatts and was 1.75 inches in diameter and 4.25 inches long.

The third-generation production laser package design is currently being tested. This design integrates the laser resonator into the diode package and reduces the length of the complete laser to 1.50 inches long, which still fits into a 1.75-inch diameter tube, and maintains the 150-milliwatt power output in a "hermetically sealable" package.

The physics, engineering, and design work on these improvements can be extended up to the five-watt level. They are working with their technology base to provide the most compact and efficient laser resonator package available. This effort will result in a package that is easily integrated into many shipboard systems.

The original application was designed to distract and disorient potential suspects at ranges beyond that of current devices being used by police officers. That system was designed to provide an eye-safe one-meter spot size at fifty feet. This was based on a spot size that an officer would be comfortable using to cover a single suspect, at the fifty-foot range, and would provide him an "optical wall" between him and the suspect. Because of the potential that the unit would be used at closer ranges than fifty feet, LE Systems designed the Laser Dazzler to be eye-safe at the aperture.

The first application was to provide a larger beam at closer ranges for use by prison personnel for cell extractions. In this case, LE Systems designed an adjustable beam expander that could be either set at the one-meter-at-fifty-foot setting or moved to provide a six-foot-diameter beam at twelve feet.

On the opposite end of the spectrum from the design that would be used for cell extractions was a long-range designator, for use by military snipers. In that case, the Laser Dazzler would be integrated into and/or onto a sniper rifle and would not be used at close ranges. For this application the beam expander would be removed and replaced with optics that would minimize the natural beam divergence of the laser source.[21]

ACOUSTICS

Scientific Applications & Research Associates (SARA), located in Huntington Beach, California, is currently the leader in development of acoustic non-lethal weapons in the United States. The CEO, Dr. Parviz Parhami, lured the current president, a Mr. Jim Wes, and several leading scientists and engineers away from major aerospace companies and the not-for-profit research establishment, included among them Mr. John Dering and Dr. Ned Patton. These people were looking for an environment in which they could explore new ideas and challenge commonly held notions in ways that were difficult, if not impossible, to implement within the structure of the large bureaucracies of the aerospace industry.

After several years of investigation into the effects of acoustic weapons, and Dering's work replicating what German scientists were doing in World War II with infrasound, SARA has come to conclusions about what acoustic weapons can and cannot do. Acoustics, applied in the sense of a less than lethal weapon, can startle and surprise a target subject, may make him angry, can disrupt his communications and activities, warn him that he is not welcome in the vicinity, and even cause him ear pain. However, SARA has discovered after years of research in this area that acoustics do not incapacitate, freeze, or stun a target subject.

There are, however, instances where incapacitation of a target subject may not be the safest approach. For instance, in the case of protecting Navy ships from attacks by terrorists, such as in the USS *Cole* incident, incapacitating the pilot of the incoming small craft would not have prevented the attack. The small boat would most probably have had enough momentum to carry it close enough to the *Cole* to still do significant damage, even if the pilot of the boat was incapacitated. Also, if the intruding small craft does not have hostile intent, incapacitating the pilot of the craft would not allow him to turn away from the ship. To deal with this problem in particular, SARA has developed a device called the Determinator. This device can determine the intent of an oncoming small craft by illuminating the craft with intense acoustics that are designed to warn the less committed that they are not welcome near the Navy ship. If the intruder persists to the point that he is doing himself ear damage, then the Navy has "determined" his intent and can deal with him in more traditional ways.

Denying personnel access to a facility or an area without killing them has also been a long-standing problem in urban warfare. Starting with the idea of using intense audible acoustics to distract a subject, SARA has added an intense intermittent light source and a malodorant to a high-pitched whistle in their Multi-Sensory Distraction Device, or MSDD. This soda can–sized device can overwhelm the senses of individuals who are in a facility or an area where the military or police forces want to deny access. The MSDD is thrown or launched into the facility or area, where it emits intense sound, a bright blinking light, and a very bad odor. The "friendly" forces can, with the proper protective gear, enter the area and take control. In addition, as with the Determinator, if the target subjects are prepared for the MSDD attack, then their intent has been "determined" to be hostile and they can be shot.

American Technology Corporation in San Diego, California, has developed a lightweight, high output acoustic weapon platform capable of radiating sound pressure levels in excess of 155dB. Successful demonstrations of prototypes were conducted at Camp Pendleton, Norfolk, San Diego, Newark Port Authority, and the Pentagon. Small-scale prototypes, appropriate for crowd control and area denial, have been produced. They are thirty-three inches in diameter, five inches thick, and weigh forty pounds including the integrated signal processing and amplification. It is man-portable but best suited for fixed land, vessel, or vehicle installation.

Since the weapon's bandwidth and directivity can support long-range hailing, several units are being deployed on shore and aboard vessels for communication with approaching craft in order to enforce naval protection zone (NPZ) perimeters. The hailing capability and intelligibility have been demonstrated over water to a distance of 500 yards. The advantages of the ATC acoustic system are its high power-to-weight ratio, scalability to smaller and larger units, and reasonable cost. Highly directional, the acoustic beam can be comfortably operated while personnel are beside or behind the device.

For shipboard use, plastic materials and the flat form factor translate to a relatively low radar return signature. In the small size mentioned previously, the system is capable of 120 dB levels at more than fifty yards, and the output scales up rapidly as the cross-sectional area is increased. The bandwidth and output are high enough to transmit highly varied audio signal envelopes in pulsed or continuous modes, including mixed tones. ATC calls their program and NLW products "HIDA™" for High Intensity Directional Acoustics, whether configured for hailing or weapon tones.

HIGH-POWER MICROWAVE

High-power microwave technology continues to have mystical qualities attributed to it. That is driven in part by the overclassification of many HPM programs. The ADS is only one of several advances that have occurred. While that was declassified so that more extensive human testing could be done, other programs remain shrouded. However, Secretary of Defense Donald Rumsfeld has indicated that details of some new HPM weapons may be released in the near future.

Size and power have been two of the key issues plaguing these weapons. Another problem is erratic test results. Specifically, identical targets exposed to the same level of energy would demonstrate a wide variance in susceptibility. Commanders want a high degree of confidence that when a weapon is used the effects can be counted on. In this case, they want to know that the electronic components will be severely damaged or destroyed.

The operational missions are quite varied, and targets include any machine or system that uses electrical parts. Communications systems, vehicles, boats, weapons guidance, and sensors are all vulnerable. Hardening can be accomplished but at added expense, and civilian equipment is inherently soft.

One of the major challenges in the application of HPM systems (such as for vehicle stopping) is the challenge of pointing the beam while the target and/or the platform is moving. Specifically, existing and emergent HPM sources and their support equipment are quite large, so putting them on a rapidly "pointable" gimbal system is impractical. One option would be to have the source off the gimbal, then route the energy onto the gimbal/antenna/director through some flexible coupling. Unfortunately, such couplings cannot tolerate the hundreds of megawatts or more power demanded by vehicle stopping and other of the highest power applications. Some systems designers have opted to simply point the entire platform (as in self-propelled howitzers or WW II–era main battle tanks).

Scientific Applications & Research Associates (SARA Inc.) has developed, proven, and patented a rapidly steerable HPM antenna. The antenna is being developed as part of the U.S. Army research lab's HPM vehicle-stopping program and will be wed with the Army source in the coming year. The antenna uses polarizing and reflective elements to first focus (high-gain) the beam, then point it through ± 90 degrees in azimuth and 70 degrees of travel in elevation. The antenna has been lab-measured to provide ~ 30 dB of gain and retain the gain within 2 dB even at the extremes of off-axis pointing. With such an antenna a wide variety of targeting and systems design choices will be provided to the designer and warfighter.

While the United States has been working diligently on developing HPM weapons, they are not alone. Several other countries also continue to make progress but are equally cloistered about their results. We can be sure that power is increasing while these weapons are getting smaller. The first notice that they have been used will probably happen when it is reported that sensitive electrical systems have suddenly burned out. Of greatest concern should be the vulnerability of our infrastructure to such attacks. Undoubtedly Homeland Security is addressing that.

RUNNING GEAR ENTANGLEMENT SYSTEM (RGES)

One concept that has emerged very rapidly is the small boat-stopping Running Gear Entanglement System. Since fast boats are a problem, the U.S. Coast Guard joined with the Navy to develop a system that can snare these planning hull craft. A metal wire, about eleven millimeters in diameter, supported by flotation equipment, is placed in front of the wanted boat. The wire can be dropped from planes or helicopters or launched with a compressed air mechanism. The RGES can bring a boat traveling at fifty knots to an abrupt halt. The boat may sustain damage from the impact, but it will stop.

EXHAUST STACK BLOCKER

Intercepting ships on the high seas has become increasingly important. For stopping larger, displacement-hull shipping, a new approach has been taken. Being tested is a 200-pound device that can be dropped into the smoke stacks of these oceangoing vessels. Once it is in place, foam is dispersed that clogs the exhaust system, causing 1,600-HP diesel engines to quit.

PORTABLE VEHICLE ARREST BARRIER (PVAB)

First mentioned in *Future War* as Speed Bump, this system has gone from research and development into production. The PVAB is an extremely strong net that can be remotely raised to stop uncooperative vehicles.

JOINT NON-LETHAL WEAPONS DIRECTORATE (JNLWD)

Formed only a few years ago to oversee the development and implementation of non-lethal weapons, the JNLWD has matured significantly. Given birth by Col. Andy Mazzara, USMC (Ret.), they quickly got basic equipment into the hands of troops deployed on peace support missions around the world.

The second director, the energetic Col. George Fenton, sold NLW "ice cream" to commanders here and abroad. He viewed himself as an ice-cream salesman, a person who knew his product was needed and had the stamina to convince some pretty hard-nosed senior officers that they could help. Colonel Fenton also saw to it that the program became formalized to ensure their research would meet field requirements.

INDUSTRY

Among the important changes that have taken place in the past few years is the involvement of major industry in the development of non-lethal weapons. At the Non-Lethal Defense IV conference in 2000 was the first time that major defense contractors began to participate in the dialogue. Now they have become full players in the field.

Of course small companies are still important. And I found a unique one in Brazil. On a warm winter day a small pond is graced with a pair of black swans while peacocks strut the grounds and Canadian geese paddle about. This backdrop seems most unlikely, but here, nestled in the forested hills north of Rio, is a unique arms company. Surrounded by plush tropical vegetation and providing refuge for exotic animals, Condor is the only company in the world to have as a product focus the introduction of non-lethal weapons for both military and law enforcement applications. The other companies developing these weapons generally manufacture a narrow, technology-driven selection of products such as electronic stun guns or pepper spray.

The father and son management team of Carlos and Frederico Aguiar has embraced non-lethal concepts both altruistically and pragmatically. Both businessmen with law degrees, they have made great strides in influencing senior law enforcement and military officials in Brazil and throughout South America. Given the prevalence of social unrest in the region, they are keenly interested in providing alternatives in use of force, ones that minimize the potential for continuation of the cycle of retribution. Therefore, they successfully transitioned a fledgling pyrotechnics company that distributed flares and smoke-signaling devices into a thriving business that employs over 200 people and is dedicated primarily to non-lethal weapons.

The Aguiar family appear to be among the first to recognize the viability of a company that markets a variety of non-lethal weapons. A global traveler, Frederico is constantly searching for new technologies that can be added to Condor's product line. Environmentally conscious, the Aguiars have demonstrated that weapons manufacturing can be integrated effectively with both nature and employees in mind. In fact, when you enter the grounds you find the fifty-two buildings dispersed in a mode more befitting a theme park than a munitions facility. With time, maybe others will adopt these characteristics.

In yet another coming-of-age event for non-lethal weapons, in a short time there will be a person with a Ph.D. in the field. Toby Feakin has been studying at Bradford University in Leeds, UK, and working on the concepts and applications of non-lethal weapons for several years. His studies have provided a unique look into the use of these weapons in several countries.

GLOSSARY OF OC TERMS

Capsaicin (I or II)

A colorless, crystalline, bitter compound present in capsicum. The amount varies depending on the species and crop year of pepper. Heat level is equal to that of vanillyl pelargonamide. Both capsaicin I and capsaicin II (VP) are the hottest of the capsaicinoids.

Capsaicinoids

The pungent components of peppers. They encompass at least six major compounds: capsaicin I, dihydrocapsaicin, nordihydrocapsaicin, homodihydrocapsaicin, homocapsaicin, and vanillyl pelargonamide (capsaicin II).

Capsicum

Common peppers.

HPLC and GC/MS

High-Performance Liquid Chromotography and Gas Chromotography/Mass Spectrometry are two of the standard chemical analysis methods to scientifically determine the concentration of compounds such as capsaicin and vanillyl pelargonamide in chemical agent compounds and sprays.

Oleoresin capsicum

The extract of the dried ripe fruits of peppers. It contains a complex mixture of essential oils, waxes, dried colored organic cellulose materials, and several capsaicinoids.

Scoville Heat Units (SHUs)

SHU testing is a "tongue" tasting to determine the heat level of the spice by a panel of five individuals. The test is not accurate since it depends on the individual taste sensitivity, which changes from person to person and does not measure the actual chemical percentage within the product.

Vanillyl Pelargonamide (VP)

Discovered in 1911, VP is a newly purified natural compound found in pepper plants. It is as hot as capsaicin I but can be extracted in purified form from other naturally occurring sources.

CHAPTER TWO

YOU CAN RUN, BUT YOU CAN'T HIDE

MANY INEXPERIENCED MILITARY OBSERVERS focus their attention on the destructiveness of new weapons. While awesome power can be delivered, sharpening the point of the spear is really secondary to locating the enemy. For decades *war games* have proven that it is more important to improve the military's sensor systems and command and control processes than it is to increase firepower. Once an adversary is located, we know how to destroy him. This philosophy has driven the research and development budget.

Hollywood has done much to distort the public's understanding of the capabilities of modern sensors. Many films depict satellites with near mystical powers. It seems that instantly the National Reconnaissance Office (NRO) can maneuver their assets to any specified area of the world and observe the most minute details. In the movie version of Tom Clancy's *Clear and Present Danger,* operations officers located in CIA headquarters in Langley, Virginia, watched as orbiting thermal sensors clearly showed the actions of each individual raider and determined whether any of the suspects remained alive.[1] In *Behind Enemy Lines,* Gene Hackman, playing a U.S. Navy admiral on board an aircraft carrier, USS *Carl Vinson,* watched as such sensors followed Owen Wilson, portraying a downed pilot, as he was pursued at close quarter across hostile territory. Eventually Wilson escapes when he accidentally falls into a mass gravesite and is covered by the decaying bodies. The partisans seem to be standing over him and yet inexplicably turn and leave without locating the pilot. Hackman, observing the body heat of the living, does not have the same view as the humans who are detecting based on the visual spectrum. At least in this film the satellites move on and only provide coverage while they are in proximity above that location.[2]

It is not just satellites that conjure up visions of intrusion of privacy. Evil government forces chased Will Smith when he inadvertently came into possession of a sensitive computer disk in *Enemy of the State.* To spice things up, moviemakers provided the audience with tantalizing images of sexual intercourse occurring inside a brick building. The detailed remote sensing took place from a vehicle parked on the street below.[3]

Such capabilities may be great for Hollywood, and the intelligence community wished they had them. Our sensors are good but not that good. And there are ways to fool them. During the air campaign against Serbia in 2000 pilots were returning with damage reports indicating large numbers of targets were destroyed in their attacks. NATO battle damage assessments listed as destroyed ninety-three tanks, 153 armored personnel carriers, and 389 artillery pieces.[4] According to then Secretary of Defense William Cohen, these losses were a crippling blow and constituted destruction of 50 percent of Serb artillery and one-third of their armored forces.[5]

Yet in postcombat analysis it was determined that the Serbian military had been extremely successful in their use of decoys. The estimates of damage are still arguable. The lowest numbers recorded were in *Newsweek,* which reported only fourteen tanks, eighteen armored personnel carriers, and twenty artillery tubes destroyed during the twelve-week bombing campaign. Many of the targets reported as damaged or destroyed were simple cardboard or wooden mock-ups designed to look like a major weapon system. The Serbs also developed means to make a target look as if it had been destroyed when it had not been hit by piling debris or painting the area to depict damage. As a result, the Serbs often successfully caused us to expend very expensive precision munitions against targets of no appreciable value. We also did not strike targets thinking them to have been previously damaged. The after-action reports were very critical of the political processes under which the war was fought and noted the need to improve the all-weather strike capability of the Air Force.[6]

From that analysis one of the key lessons learned was the value of *eyes on target.* In Afghanistan the targeting frequently was directed by special operations forces who were on the ground and as close to the objective area as they could get. While very effective, these tactics greatly increase the vulnerability of those forces and are practical only when air supremacy is assured and adequate fire support is constantly available. There are many situations in which leaders must weigh the value of close identification of targets against the possibilities of having those troops killed or captured.

WHAT HEAVEN KNOWS

Dominating the space dimension of military operations is key to winning all future conflicts. Once the domain of specialized Air Force personnel, the capabilities provided by space assets now affect every U.S. combatant. They permeate all aspects of our fighting capability from reconnaissance and surveillance to command and control, navigation, planning, and, finally, guiding the destructive force of sophisticated weapons. While some movie versions of our surveillance capabilities might be exaggerated, for the past three decades there have been constant improvements, and more are on the way. To paraphrase the Air Force motto, when it comes to space, *no one comes close*. As we shall explore later, there are others emerging in the space race and the odds are likely to change. But for the near future, the United States will continue to dominate.

The *eyes in the sky* have served us well. They contain a vast array of sensor systems that are constantly being improved. Some in low Earth orbit, 200 to 600 miles high, provide imaging, communications, and weather data. Navigational satellites, such as the Global Positioning System (GPS), range from 6,000 to 12,000 miles up. And finally there are geosynchronous satellites as far out as 22,000 miles that have twenty-four-hour orbits that keep them over a fixed location on the Earth. All are integrated into a complex set of intelligence and command and control systems that enable warfighters at every level of command.[7]

Strategically, the Defense Support Program (DSP) holds the high ground. At 22,000 miles the DSP satellites can observe about half of the planet. From the first publicly announced launching in the 1970s, they have been scanning the Earth with infrared (IR) sensors for the signatures of nuclear launches. They look for the flare of ballistic missiles and can detect nuclear detonations.

However, as the critical mobile target problem became more serious, it is believed, these satellites were modified to be able to detect launches of medium-range missiles as well as ICBMs. Most readers will remember the great SCUD hunts that took place during Desert Storm when Iraq fired missiles at both Saudi Arabia and Israel. While the actual SCUDs were not militarily effective, they had substantial psychological and diplomatic impact. In response we were required to divert considerable special operations forces and air force assets to hunt for launchers hidden in the desert and capable of moving periodically.

The DSP satellites are quite large platforms. They are reported to be over thirty-two feet in length and twenty-two feet in diameter with solar panels extended. Weight is estimated to be about 5,000 pounds.[8] Some believe that the DSP systems can distinguish between U.S./Russian weapons and the detonations from Third World nuclear powers such as India and Pakistan. Such discrimination could be critical in the analysis should a nuclear detonation take place.

But in general the DSP satellites are old and need to be replaced, and so a new generation of early warning system is on the way, and it comes in two major subsystems. The Space-Based Infrared Systems (SBIRS) will be formed by two groups of satellites, or constellations, as they are known in intelligence circles. The SBIRS-High version with four geosynchronous birds will search for missile launches, which it can accomplish more rapidly and accurately than the DSP system.[9]

The SBIRS-Low version, a constellation of twenty, is designed to track missiles from low Earth orbit. Importantly, Low can discriminate between real warheads and decoys. Ever since other countries developed multiple-warhead technology, there has been a scramble to be able to quickly identify the difference. Employing a combination of both scanning and staring IR sensors, SBIRS-Low will be able to do that job.

A warning component such as SBIRS is essential for support of any space-based missile defense system. Looking down from 60 to 300 miles up, this system will be able to have stereoscopic viewing of missiles in flight providing precise location, velocity, and acceleration data. This accuracy will provide the defender much faster reaction time and permit intercepts much earlier and extend the defended boundaries three to five times. In addition, the tracking station can determine options for destruction of the incoming missile so that it can be destroyed over unpopulated areas.[10]

The speed of detection also will establish the location of launchers, allowing them to be attacked before any additional missiles can be launched. An integral part of the future National Missile Defense system, SBIRS is called the *Eyes of the Viper*. Designed for war, the system will be active during peacetime. Since it covers the globe, SBIRS will be constantly collecting data from every missile test conducted anywhere on Earth. The data on signatures and trajectories will be used to update the database and better enable the satellites to identify missile launches, then hand them off to weapons systems for destruction.

Also in geosynchronous orbit are the NOAA (National Oceanic and Atmospheric Administration) weather satellites. While they provide weather data for civilian use, they are critical for supporting military operations. In today's era of quick raids and supporting air strikes, planners need to know exactly what to expect as far in advance as possible.

In addition to the DSP system several others serve the intelligence community. Spy stories are filled with tales of electronic intelligence or, as they are sometimes known, constellations' ELINT capabilities. These are comprised of both ground-based and space-based collection facilities. They are very good at collecting signals. However, there is so much data being collected, sorting is the real problem. There are also various image intelligence, or IMINT, systems. Some of these are photo-optic satellites. In the beginning they would circle the Earth, take pictures, and then drop them by parachute. A retrieval plane would catch the containers in midair and return the film for processing.[11]

Now there are electro-optical/infrared capabilities that provide photographic imagery on a full spectrum and can see into the infrared. These systems can photograph what the eye can see, albeit somewhat enhanced, and spot heat. The imager does not penetrate heavy clouds or foliage but can locate engines that have been running and camps. In fact, during Enduring Freedom the onset of winter was welcomed. It was known that the heat signatures from fires or vents in caves would stand out from the natural environment.[12]

Exactly how good these EO/IR sensors are remains classified. There are decades-old stories about reading license plates in Russia. Robert Roy Britt of Space.com notes that cameras from low Earth orbit can spot items as small as four inches. He also says the Key Hole satellites can read headlines while in orbits well in excess of 200 miles high.[13] By contrast, he credits the civilian Ikonos imaging satellite with recognition of objects over three feet across.

The third imaging capability is provided through use of radar. High-energy pulses are reflected by the Earth's surface and captured. The mature Synthetic Aperture Radar (SAR) provides detailed images in day or night. The waveforms allow the radar to penetrate clouds and dust that block visual and infrared signals. Doppler radar identifies movement. It is good for locating planes and ships, while the ground-moving target indicators can detect the motion of vehicles.[14]

For many years any information about the Key Hole series of satellites was highly classified, as was the organization that was responsible for them. Until the late 1990s, very few people knew of the existence of the NRO. Now nearly everyone has seen declassified pictures such as those used by then Secretary of Defense Cohen and Chairman of the Joint Chiefs of Staff Hugh Shelton that depicted the al Qaeda training camp at Zhawar Kili, Afghanistan. They were from the older KH-11 series and used to support the 1998 cruise missile attack.

Later came the KH-12 series of launches also known as Improved Crystal. The satellite is slightly larger than the Hubble Space Telescope. Improved Crystal has newer sensors including image intensifiers for improved night-time images, optics that operate in the visible, near-infrared, and thermal infrared regions. It has the capability of onboard digital enhancement of the photos that are transmitted instantaneously back to the analysts. The sun-synchronous KH-12 travels low and fast and passes over a given site at the same time each day. This allows analysts the ability to compare changes from one day to the next. A big advantage is that it carries much more fuel than the KH-11 with a total weight of about eighteen tons. This fuel enables the satellite to be repositioned to areas of interest and may add to its antisatellite evasion capability.[15] It should be noted that these satellites are not cheap. In fact, considering that the cost to orbit per pound is about ten thousand dollars, the launch cost alone for an eighteen-ton object would be about $360 million.

To augment the Key Hole satellites another series, called Lacrosse, began development in 1976 under then Director of Central Intelligence George Bush. The first launch from the Space Shuttle was on December 2, 1988. The remainder were boosted from Vandenberg Air Force Base on Titan 4 missiles. The fourth launch was on October 17, 2000, under the name Onyx. These satellites with large antennae beam cloud-penetrating micro-wave energy to the ground and read the reflected signals. These sensors have limited ground- and water-penetrating capability and can locate bunkers up to three meters deep and submarines at periscope depth. Onyx also uses SAR technology but seems limited to identification of objects three feet across.[16]

There is a new constellation of NRO satellites that began to be launched in 1999. Various names have been associated with this system. They include KH-13, 8X, and EIS (Enhanced Imaging System). Built by Boeing, the NRO has repeatedly announced successful launches but made no comments about capabilities.[17] It is known that they are in highly elliptical orbits. The October 11, 2001, launch was calculated to vary between an apogee of 20,246 miles and a perigee of just under 148 miles. The high-quality sensors can accommodate the long distances, and the geosynchronous orbit allows continuous battlefield surveillance, something sorely missed in Desert Storm. It is considered likely that there will be twenty-four of these multifunctional satellites that will provide coverage of all points on Earth every fifteen minutes.[18] The imaging sensors are undoubtedly far better than those of the KH-12.

While sensors, image enhancement, and signal processing continue to improve, the military is working on other technologies to increase the effectiveness of satellite systems. Due to their size, repositioning them requires a lot of fuel. To enable the movement, much of the volume of the spy satellites is taken up by fuel containers. The Defense Advanced Research Projects Agency (DARPA) has initiated Orbital Express, a program to test the feasibility of refueling the satellites with a tanker from the Space Shuttle. If this is successful the satellites could be smaller and routinely serviced by the external source.

One concern for military leaders is the availability of commercial satellite imagery. In Desert Storm, Operation Left Hook could not have been executed with surprise if Saddam Hussein had had access to information about the coalition forces positioning supplies far to the west. While the civilian images are not as good as those of the intelligence community, they are good enough to provide warning to future adversaries. It has been agreed that antisatellite systems will not be used.

That may have to change. An alternative is to simply buy exclusive rights to all imagery under certain specified conditions. More complex is the issue of shared capabilities, in which several nations use the services of commercial satellites. The time may come when we choose to use weapons to destroy or disable adversarial or neutral satellites. That decision must be made with great care, as our own commercial satellites will be vulnerable to counterattacks.

Potential adversaries should be aware that our satellite reconnaissance program is improving. Our ability to detect targets at night and under adverse weather conditions will continue to advance. The next-generation satellites being launched now will provide substantially enhanced capabilities for ground and air combat commanders.

Of course satellites are not alone in peering down from the heavens. The development of the U-2 spy plane more than sixty years ago ushered in a new era of aerial observation. The U-2 is still flying today, and it continues to be upgraded. However, it was the May 1, 1960, downing of the Black Lady with CIA pilot Francis Gary Powers near Svedlovsk, USSR, that pointed out the most serious shortcoming of manned aerial reconnaissance missions—the pilot could be captured. That U-2 incident had huge diplomatic implications and was an international embarrassment. When Nikita

Khrushchev demanded that President Eisenhower publicly apologize it caused the breakup of the Paris Summit. Following that, Powers was tried and sent to jail. He served twenty-one months before he was traded for Soviet spy Col. Rudolph Abel.

Concern for vulnerability of pilots and advances in remote control technology have led to the development of new fleets of unmanned aerial vehicles (UAVs). They are defined as "powered aerial vehicles sustained in flight by aerodynamic lift over most of their flight path and guided without an onboard crew."[19] While the definition may seem a bit arcane, it describes the process while eliminating certain technologies. In fact, aerial observation of the battlefield began with balloons in the 1800s.

The path to today's UAVs was not easy. True, model planes have been around for a long time, but developing remote-controlled aircraft capable of accomplishing military missions cost many millions of dollars. They have several modes of operation, including preplanned routes, remotely piloted, and updated autonomous flight patterns.

UAVs have become a mainstay of reconnaissance and are known for their ability to conduct missions that are *dirty, dull,* and *dangerous.* Unmanned, they can be placed in moderate-to-high-risk situations without fear of loss of life. They make excellent sensors for use in potentially contaminated areas. Given the rising concerns about chemical and biological weapons, UAVs can be flown directly into those areas to collect samples. With existing miniaturization of sensors, they can even do onboard analysis and relay the information to their control centers. Humans become bored when involved with highly repetitious functions, but automated sensors do not. Therefore, UAVs make superb sentries that don't tire easily.

While not the first UAV, the best-known is the RQ-1 Predator. It was designed for reconnaissance and is a medium-altitude long-endurance system. The Air Force contends the Predator is a system, not just an aircraft. Each operational system has four UAVs, a ground control station, and a Predator Primary Satellite Link. A crew of fifty-five people is assigned to conduct twenty-four-hour operations. Powered by a Rotax 914 engine, the Predator has a range of 454 miles and can carry 450 pounds. The system reportedly cost $40 million.[20]

One pilot and two sensor operators fly the Predator via a line-of-sight data link. When the UAV is beyond line of sight, commands are given through a Ku-Band satellite link. The color nose camera is used primarily for flight control. It has a daylight television camera supported by a variable-aperture infrared camera for low-light level conditions. The Predator's SAR allows it to image through smoke and clouds in still frames.

Predator was a successful example of the congressionally mandated Advanced Concept Technology Demonstration (ACTD) program, which is designed to quickly field promising technologies. It was demonstrated as a proof-of-concept in the spring of 1995 and by that summer was deployed for use in Bosnia. It has been employed on missions ever since that time.[21]

There are frustrations inherent with remote operations. One such moment came when a loitering Predator observed the ambush of a special operations helicopter that was attempting to land a team in the Afghan mountains. Struck by heavy fire, including an RPG round, Petty Officer 1st Class Neil Roberts, a SEAL, was physically knocked off the CH-47. The pilot of the crippled craft struggled to maintain control and, luckily, set it down some distance away.

Roberts was stunned but able to continue to fight. Surrounded and out of ammunition, he was captured by the Taliban fighters. The Predator could only watch as his captors took him away. When a ground rescue party finally returned they found the body of the executed SEAL.[22]

Both the Air Force and CIA have operational Predators. The CIA had asked to arm the UAV so that targets could be immediately engaged. The General Atomic designers were concerned about whether the UAV could withstand the launch of a Hellfire missile. In February 2001 the Predator variant successfully demonstrated that the laser-guided Hellfire-C was viable. In a series of tests designed to determine if Predator could be used in an antiarmor role, it destroyed twelve of sixteen targets and assumed a new combat role.[23]

The armed Predator has been deployed successfully in other areas to counter terrorism. On Sunday, November 3, 2002, CIA operatives in Yemen learned the location of Qaed Salim Sinan al-Harethi. A known terrorist, suspected in the bombing of the USS *Cole* in December 2000, he was high up on the most wanted list. Acting on information probably attained through time-honored bribery, al-Harethi and several others were spotted driving an SUV in a deserted area. The group probably never knew what hit them, as their vehicle disintegrated so that the only recognizable feature was a spring of the suspension system.

Among those identified with al-Harethi was Kemal Darwish, an American who was suspected of being the recruiter of the terrorist support cell arrested in Buffalo, New York. While questioned by some legal analysts, this strike sent a clear message that the United States would take action against terrorists wherever they were found. The international aspect of this operation is seen in that a U.S. source flew the Predator from a base in Djibouti to attack the target in Yemen.

Following on the success of Predator, its big brother could not be far behind. Global Hawk, with a wingspan of 116 feet and range of over 12,000 miles, began development in 1995 as another ACTD. In June of 1999 it was tested by the Joint Forces Command, and by April of 2000 it completed the first transoceanic flight to Europe. Then in May 2000 Global Hawk supported a Navy Carrier Group and an Amphibious Ready Group/Marine Expeditionary Unit in a joint land-sea operation.

Demonstrating trans-Pacific capability, in April 2001 Global Hawk set the UAV endurance record with a 7,500-mile trip to Australia.[24] The first operational deployment was to Afghanistan on October 7, 2001, in search of Taliban and al Qaeda forces.[25]

Global Hawk can fly to 65,000 feet, with an endurance time of up to thirty-six hours. From that altitude it can conduct surveillance of an area the size of Illinois, or about forty thousand square miles. With a 2,000 pound-payload of high-resolution sensors it will provide field commanders with information dominance throughout the battle-space.[26] The sensor suite includes SAR with a one-foot-spot mode of detection and a four-knot-minimum moving target indicator that provides both speed and direction of travel. These combined with the IR and EO sensors provide an unprecedented ability to monitor large areas and distribute the data in near real time.[27]

There is a need for UAVs at all levels. On the horizon are small, inexpensive ones that can be controlled by front line commanders and provide information about what is over the next ridge or around the corner. The exigency for such capability may actually have led to the demise of a program designed to provide the infantry with a smart, non-line-of-sight antitank missile. For decades the Army was developing a Fiber-Optic Guided Missile (FOG-M). It had an excellent test record for finding and destroying tanks. Vertically launched, the FOG-M would be flown downrange while transmitting a picture back to the remote-controlled pilot. The wide-angle lens allowed the operator to look for potential targets. When one was spotted, it would be selected and the gunner would fly the FOG-M directly into the target. Success was demonstrated against various armored vehicles and even flying helicopters.

Unfortunately for the FOG-M program, these capabilities were selected for use in a war game. Lacking UAVs at the time, field commanders frequently selected to fire a FOG-M as a means of surveillance of the battlefield. Once fired, the missiles could not be recalled. Therefore, if no targets were observed, the FOG-M flew until it ran out of fuel or cable. Controllers graded these missions as failure of the missile. When word reached Congress that FOG-M had such a low kill rate, they decided to cancel the program. The reality was that the missile performed the designated task quite well. Had a small UAV been available at that time, both programs would be alive today.

What those commanders wanted is now on the way. The intelligence problem for front line troops has always been getting access to the information that is useful to them. The expensive systems we have mentioned will be controlled at very high levels of command. The troops need something they can deploy quickly and get immediate feedback. To be useful it cannot require special processing stations and long runways from which to take off.

DARPA is working on an intermediate solution. The prototype will be twenty-two inches tall and weigh less than five pounds. The mission will be to dart behind an enemy location, take photographs, and get back home without human intervention. The technical challenge has been to miniaturize the parts for a large aircraft so that they can operate in the UAVs. This system has vertical takeoff and landing, so it can be deployed from most any location. Still, by micro-UAV standards, this is pretty large.

Work has been done to make micro-UAVs that are only a few inches long. Of course that means that the sensor packages must be extremely light. Cameras that weigh only a few ounces are proposed. Power is the long pole in the project. Batteries take up weight and space. Reducing the size of batteries also reduces the power available. The use of microwaves for powering these craft has been demonstrated. This requires that a beam be projected to the craft to keep it flying. Of course there could be another airborne platform that sends the microwave beam down to the UAV. That would allow the micro-UAV to fly behind buildings, hills, or trees without a large auxiliary power source.

Unlike larger UAVS that require long runways, their micro cousins may be launched in a variety of modes. Some will lift vertically like miniature helicopters, while others may be thrown into the air by an operator. To reach out a bit farther they may be tube-launched. The compact micro-UAV would have its wings folded inside a canister. A small rocket would project the craft to the target where the wings would deploy, and the UAV would loiter above the search area, transmitting pictures back to the unit commander on the ground. All of the technologies necessary to build these micro-UAVs have been developed. Remaining is the engineering task, but that is well under way.

SEE-THROUGH-WALL TECHNOLOGIES

One of the highest priorities of both military and law enforcement officers has been the ability to accurately determine what is on the other side of a wall, buried underground, or hidden in the trees. While this is at the top of the list, there have been many ideas put forth with little success. To this date the best way to see what is behind a wall is to drill small holes and insert optical devices. Many of these are very small and drilling operations relatively quiet. Of course observation is limited to the location selected.

There have been numerous attempts to develop electromagnetic sensors that can penetrate various kinds of materials. In general, these systems require operators to place the sensors either on the wall or in very close proximity to it. Relatively large, these comprise sensors are weighty items that must be carried by someone, and it will be some time before they arrive on the battlefield.

However, there have been significant advances in several areas. Millimeter waves, radar, X rays, and radios are being used in new ways. Also being explored are optical, thermal, magnetic, and acoustic sensors, all capable of detecting targets or contraband items. At the heart of many of these systems are neural nets that allow computers to analyze for minute differences in signature characteristics. Without modern computational capabilities, many of these advanced systems would not be effective.

Since 9/11 there has been increased emphasis on developing an ability to detect weapons being carried by people. Actually, the police and military have been requesting these items for many years. Now that they are a higher priority for homeland defense purposes, both research and fielding have been stepped up.

Airport security has led the way. Several body-scanning technologies have come forward. There was an uproar in the media when the first body scan images were publicly released. There, for all to see, was every anatomical detail of the person's body being scanned. Many believed that amount of detail went too far when the objective was to find weapons.

Millimeter wave technology is very promising for penetrating walls. These extremely short waves, 30 to 300 gigahertz, can be transmitted through a wall and echo off objects on the other side. These echoes are captured and through signal processing converted into usable images. It is claimed that in tests conducted on a system built by Millivision, individual termites could be spotted inside a piece of wood.[28] Millimeter waves at these frequencies cannot penetrate human skin and are thus safe for scanning people. They do allow an operator to determine the difference between the human body and metal, plastic, or ceramic weapons that someone might attempt to smuggle.[29]

If motion detection is the objective, then Time Domain's RadarVision 2000 should meet that requirement. Time Domain states that they can detect motion as deep as thirty feet behind a wall and plot the object's location and rate of movement. Currently they are working on a small portable system for Army applications.[30]

FACE RECOGNITION

In the counterterrorism game, face recognition is rapidly becoming one of the critical tools available. Still far from perfect, face recognition systems can be used to screen large numbers of people very quickly. The systems work by measuring specific aspects of facial features, such as the width of the eyes and height and separation of cheekbones. Sophisticated systems that rely on neural net processing, facial recognition technology searches for key characteristics and establishes anchor points. Then the systems connect the pixels forming a series of triangles. The systems search for abrupt changes between pixels indicating a facial feature. These are converted to a number comprised of 672 ones and zeros that identify the face of a given individual.[31] That information is then compared with faces in the database to determine if a match can be made.

A derivative of automated recognition systems, facial feature recognition systems have been worked on for decades. However, it has only been in the last few years that the advances were sufficient to transition from research to applications. Much of the effort was conducted under the auspices of the Department of Defense Counterdrug Technology Development Program Office.[32] The program, known as FERET, or Face Recognition Technology, was conducted at several major university and private companies, including MIT, Rutgers, the University of Southern California, and the Analytic Science Corporation. Since then there has been a concerted effort supported by the Department of Defense, Department of State, Department of Energy, and Department of Justice in which tens of millions of dollars have gone into advancing the research.

Developers believe that with the use of sufficient anchor points simple disguises, such as sunglasses or beards, will not fool the system. Others are not so sure. In the past few years several companies have marketed face recognition systems. There have been several highly publicized applications, including use at Raymond James Stadium in Tampa for the 2001 Super Bowl. Others include some shopping malls and at selected airports, where test results have been rather mixed at best.

Almost instantly there were complaints about invasion of privacy. Use on an unsuspecting crowd at the football game precipitated the initial complaints. Beating the drum most loudly was the ACLU. However, it was the results of a two-month test that was conducted at the Palm Beach International Airport that evoked their most vocal wrath. In fact, the test indicated that none of the persons targeted were detected by the system. The commentary by Barry Steinhardt of the ACLU included the remark that a system had to be perfect before it should be employed. Given today's exigent circumstances, such comments are deleterious to our national security. This is not an academic game in which we are engaged.

There are applications for face recognition that have nothing to do with criminal identification. One of the most hopeful is use in locating victims, especially children who have been kidnapped. By comparing scanned pictures from earlier childhood, face recognition systems will be able to identify children even after they have aged a number of years. Face recognition is also being used for determination of youngsters who have been exploited in child pornography.[33]

Most impressive is the distance that has been covered in face recognition technology in the past decade. Without the basic work supported by neural networks and fast computers it would not have advanced as far as it has. There is no doubt face recognition needs improvement. That will come quickly and distractions from the likes of the ACLU will not help. Militarily, face recognition has a lot to offer. We know from peace support operations in the Balkans that criminals wanted for high crimes passed through NATO-controlled checkpoints. A face recognition system would assist the troops manning those positions. The time will come in the near future when combat units will be able to use a small camera and transmit the images back to a central processing site offering near real-time positive identification. Similarly, these systems will enhance our counterterrorist capability significantly. The effort at the moment should be to establish the largest possible database from which comparisons can be made. It will be easier to get the photographs before these individuals turn to terrorism rather than after they begin hiding their faces under their masks.

DETECTION OF DECEPTION

Detection of deception has been pursued from time immemorial. Until very recently there have been no systems that fared much better than chance. After decades of use, even evidence provided by the much-vaunted polygraph has yet to become admissible in court. Too late we have learned of foreign agents who have routinely evaded detection from routine tests administered as a requirement for their admission into our most secret programs.

A case that came to light in 1988 was that of Larry Wu-Tai Chin. A career CIA analyst, Chin had already been recruited to spy for China before he joined the Agency. After being accepted on the basis of his entrance polygraph examination, he went on to pass routine examinations that were administered throughout his twenty-year career.[34] So, too, did Aldrich Ames pass repeated examinations even though he registered as *deceptive* and there were numerous flags that should have warned his CIA supervisors that something was wrong. In fact, Ames reported that his Russian handlers "laughed at his worries about polygraph exams." In their view, such exams didn't work.[35] Less known is that about thirty Cubans working for the CIA were actually double agents working for Cuba. They, too, passed their polygraphs.[36]

Other agencies have had spies in their midst who successfully evaded the polygraph. Senior FBI Agent Robert Hanssen eluded detection for many years while spying for the Russians, as did Ana Belen Montes of the Defense Intelligence Agency. Attorney General John Ashcroft alluded to part of the problem when he stated that 15 percent of the results produce false positives. That is, they wrongly identified a person as being deceptive when he was not.[37] Cleve Backster, a polygraph expert largely responsible for developing the techniques used in the field, disputes that figure and suggests the accuracy rate is actually in the mid-90s as a percentage and the larger number includes inconclusive examinations.[38]

However, the error rate is sufficiently high that it severely diminishes the credibility of the process in the eyes of many. In addition, such exams have ended careers. Chief Petty Officer Daniel King was investigated, jailed, and tried by court-martial after a "no opinion" analysis was rendered on a routine polygraph examination. The judge dropped all charges. Similarly Mark Mullah, a career FBI agent, was investigated and returned to duty. However, he never actually regained the trust that had been placed in him.[39]

It should be noted that the polygraph is a cumbersome, invasive process that relies on the interpretation of a trained operator. Anyone who has been subjected to routine polygraphs knows they are never fun. Even the short version, such as the counterintelligence examination, will take three to four hours. The basic issue is whether or not you are a spy or agent for a foreign government. The lifestyle polygraph is even worse, as the operators will ask questions about your most private matters. At one point in recent history it became clear that operators were more interested in the prurient aspects of one's life than catching spies. Thereafter, extemporaneous questions such as your favorite position for sexual intercourse were forbidden.

Law enforcement officials are desperate for a reliable and portable system. In fact, when the National Institute of Justice held a conclave to explore emerging technologies the need for a portable deception detection system was at the top of the list. Capt. Sid Heal, Chief of the Special Enforcement Bureau of the Los Angeles County Sheriff's Department, noted that they did not need a perfect system, just something better than what was currently available.[40]

Voice stress analyzers have been around for many years, and in desperation some police departments have bought them. The theory is that when a person is under stress the vocal cords will tighten and these devices can detect the change. The fact is that sometimes the systems work. However, as determined by a 1996 study by the DoD Polygraph Institute, their batting average is no better than chance.[41] Psychologically, the presence of any device can increase the probability of obtaining a confession.

As there is nothing available, police officers have become quite innovative. One often related story is that suspects would be told to place a hand around the antenna on a police car. They were told that if they lied they would receive a shock. When a question was answered, another officer would key the radio microphone placing a load on the antenna, thus providing the suspect with the promised shock. The operating assumption by many officers is that all suspects are being untruthful. As seasoned Sergeant Bobby Alcaraz, who works for Captain Heal, put it, "We are in the lying business."

Under the experienced tutelage of businessman Harry Rosen, a foreign-developed deception-detection technology was brought to the United States for further exploitation. After making an initial determination that the system did work as advertised, Rosen engaged senior scientists who formerly worked at Los Alamos National Laboratory to validate the research. A consulting mathematician, George Baker, wrote that he had examined the mathematical analysis involved in the expert system and concluded that it represented a new and unique intellectual property. That analysis resulted in the U.S. patent being issued for the personality assessment technology.[42]

The system works by capturing utterances of the human voice and comparing them against a very large database. The central database was painstakingly constructed over a period of twenty years and included over 18,000 individuals. The indicators were established through extensive testing and interviews with each research subject. Then advanced algorithms were developed to provide a response in near real time. These can be accomplished on a personal computer or laptop. The system is totally unique in that it does not rely on the content of the words but rather locates fractional properties of sound that indicate the speaker is being deceptive or intends to deceive. These minutes bits are below the level that can be consciously controlled by the person. It is important to note that the principles involved are very different from those of the voice stress analyzer systems previously mentioned.

The first market application will be under a company called Law Enforcement Agency Detection System (LEADS). Due to legal considerations and the potential for abuse, the technology will only be made available to law enforcement, intelligence, and other security agencies. While not a panacea, the technology can provide instant clues about the validity of statements made by any individual. It will then be up to the operator to explore what that person is being deceptive about. It should be noted that the system does require that a person speak a minimum of a phrase. Simple *yes* or *no* responses are insufficient for the analytical process.

What will be disconcerting to many people will be that any human utterance may be subjected to this analysis. The system is totally noninvasive and does not require the cooperation of the individuals being investigated so long as they speak. It has been applied successfully to taped police interviews and television recordings. One pertinent example is the famous comments by former president Clinton when he was denying his sexual involvement with Monica Lewinsky. Needless to say, the analysis indicates deception at several points, as he stated, "I did not . . ."[43]

The system is both content- and language-independent and requires little training to employ. Security personnel will cheer the advance, while privacy rights advocates are likely to decry LEADS as yet another invasion into human activities. Rules similar to those of wiretapping are likely to apply.

However, criminal suspects and those being interviewed by intelligence agents have a reasonable expectation that their answers are being recorded and that the investigator is attempting to ascertain the veracity of the comments. Therefore, the technology will be held to be authorized for use in such situations.

In addition, there are several circumstances under which an individual is compelled to be truthful. For instance, once in custody a person is required to accurately identify him- or herself and provide other pertinent information that does not directly relate to self-incrimination, and thus he or she remains protected under the Fifth Amendment.

While law enforcement applications obviously abound, LEADS also has military utility. As a result of Operation Enduring Freedom a considerable number of suspected hard-core Taliban and al Qaeda members were detained. Some of them were taken to Camp X-ray at the U.S. Marine base located at Guantánamo Bay, Cuba. Intelligence interviewers soon learned that these detainees had been well trained in counterinterrogation techniques, and obtaining the most basic accurate information was extremely difficult. For example, the detainees would give different names to their interrogators. Thus the government agents weren't even sure of who they had in custody, let alone anything about their past affiliations or activities. In such a case just the ability to confront the suspect when deception was indicated would be useful.

It is foreseeable that LEADS technology could play a wider role in future military operations. During peace support operations and in handling refugees many people are questioned about a wide variety of issues: Whom do they support? Where are they going? Where do the live? Where are they coming from? Are they aware of impending actions? Do they have arms? The list could be exhaustive. There is an additional problem in questioning in that usually translators are required and nuances of information can be lost. As LEADS is language-independent, it would be very helpful in assisting forces in maintaining peace and stability.

There may also be applications for LEADS in actual combat missions. It is highly likely that many future operations will be conducted in cities that have a substantial number of noncombatants. Separating them from adversarial forces will be an essential task as troops clear building after building. Envisioned is a small radio transmitter carried by an infantry squad member. Suspicious people would be asked to state their business in the conflicted area. Central processing could be done at a remote site. The response could be a simple red or green light indicating probable presence or absence of danger. Should suspects get a red light, the squad would know to handle them with caution, possibly as potential prisoners.

During Desert Storm, coalition forces had difficulty dealing with the vast number of prisoners who were attempting to surrender. Here again a LEADS device could be helpful in determining whether the individuals were intent on surrendering or possibly attacking our forces when their guard was down. One quick issue to determine would be whether the person was carrying any concealed weapons or explosives. LEADS could also be used in the event a senior officer was attempting to portray himself as an insignificant soldier of lower rank.

While LEADS deception technology is useful, the next-generation systems will be greatly advanced. It already has been demonstrated that the technology can produce a full personality profile based only on limited input. All that is required is for the person to speak thirty phrases on any subject. In less than one second the computer can display a graph that depicts the basic personality type of the person. As fast as a printer can spit out the material, a complete multidimensional personality inventory will be delivered to the operator.

Here again the potential applications are enormous. A hostage negotiator could have a short discussion with an abductor and know better how the suspect is likely to respond to various situations. So, too, detectives and other interrogators could quickly determine the best method by which to confront a suspect. It is anticipated that LEADS will be more useful in removing people from suspicion than positively identifying the guilty.

Of course both law enforcement agencies and the military rely on politicians for funding. For purchases over specified amounts law enforcement agencies and the military must obtain authorization for expenditure of funds. It is reasonable to expect many elected officials to be extremely concerned about the availability of LEADS technology. As the old saying goes, "You can tell a politician is lying when his/her lips are moving."

THOUGHT POLICE

On the horizon is another technology that may allow investigators to peer into the inner sanctum of the mind. For the past decade work has been advancing on examination of unique brain wave activity. Lawrence Farwell, of Brain Wave Science, claims to have tested more than 170 people with a

100 percent accuracy rate in determining whether individuals have been exposed to certain stimuli. For instance, do they recognize pictures of the site of a murder or words specifically related to that incident?[44]

The system works using the principles of the electroencephalograph, or EEG. Suspects must be cooperative in that the electrodes are attached to their heads. Farwell suggests that the suspect cannot simply think about another topic and avoid sending the specific EEG signals that correlate to the crime. The waveform, called P300, is fed into a computer, at which point a proprietary algorithm conducts the analysis and determines whether the subject is responding to images that are presented on a screen.

Another technique for scanning brains tracks the blood flow using a functional magnetic resonance imaging machine (fMRI). The concept is based on research that suggests it takes more effort to lie and the fMRI can measure brain functioning. Specifically, they look for increased activity in the prefrontal and premotor cortex and the anterior cingulated gyros. It is the neurons firing as they consume oxygen that indicate the person is working harder at being deceptive.[45]

These brain-scanning technologies will have utility in the war against terror. The time will come when authorities will wish to determine whether certain people have caused acts of violence or are likely to cause them. The process still requires a cooperative subject. However, one can imagine this process being used in a case such as the search for the person who mailed the anthrax-laden letters in the fall of 2001. Since the search was narrowed to a few individuals with advanced knowledge of biological warfare techniques, it would be feasible to have staff members volunteer to take the test.

CHAPTER THREE

THE LETHAL LEGACY

FUTURE WAR: NON-LETHAL WEAPONS in Twenty-first-Century Warfare introduced many readers to the concept of non-lethal weapons. To understand future conflicts it is imperative to know that lethal and non-lethal systems will be seamlessly integrated. *Future War* noted that non-lethal systems should never be employed without backup and that bluffing should not be attempted. Non-lethal weapons are a pragmatic application of force and should be used only when the political or diplomatic situation warrants use of force.

Having explored the major advances in non-lethal weapons, it is time to turn attention to their lethal counterparts. Given the breadth of emerging systems and technologies, no single book can cover every aspect of new weapons. As Secretary of Defense Donald Rumsfeld has repeatedly pointed out, the military is in a state of transformation. This has been brought about due to the change in the nature of conflict. Before embarking on the trail of advanced weapons systems, a word of caution is advisable. Just as non-lethal weapons should not be used without lethal ones, all of these traditional systems must be juxtaposed with the complex concepts that come later in the book. In the future, it will be critical for concept integration to include military force, economic considerations, perception management, and political constraints. There are no stand-alone systems.

DEATH FROM ABOVE

"As you would expect, they make a heck of a bang when they go off and the intent is to kill people." That was the comment of Gen. Peter Pace, Vice Chairman of the Joint Chiefs of Staff, when describing thermobaric bombs.[1] Before the attacks on the caves of eastern Afghanistan, these bombs were virtually unknown to the public. The term *thermobaric* is assigned to a new class of fuel-rich explosives that release energy over a longer period of time. This creates a long-duration pressure pulse when the bombs are detonated in confined spaces.[2] When these bombs were dropped, reporters at considerable distance from the targets commented on the qualitative difference of the explosion. The force of these bombs was far greater than anything they had previously experienced.

In Vietnam the U.S. Air Force dropped the first of these non-nuclear megabombs. Known as Daisy Cutters, BLU-82Bs carried a 12,600-pound warhead with slurry consisting of ammonium nitrate, aluminum powder, and polystyrene. The name came from the manner in which the bomb sliced through trees and could create helicopter landing areas with a single blast. The weapon produces an over pressure of 1,000 pounds per square inch at ground zero.[3] Later, during Desert Storm, eleven such bombs were dropped from Special Operations C-130 aircraft. In addition to blast, their psychological effect was enormous.

However, contrary to many news reports, the earlier versions were not fuel-air weapons. Those came along during Enduring Freedom. The U.S. military has long been concerned with deeply buried, hardened targets. Missile launchers and command bunkers hidden underground were very hard to successfully attack. The development of bombs such as the BLU-118/B provided a big advantage over the earlier systems.

Accelerated by events of 9/11, the BLU-82/B was put on a fast track to finish development and testing. On December 14, 2001, the first successful flight was conducted in the Nevada desert. The warhead was delivered by a guided bomb and demonstrated that it could hit a pinpoint target and deliver devastating effects. Unlike the predecessors that had to be dropped from cargo aircraft, this new bomb could be carried by an F-15E and would find its way to the target by either laser tracking or satellite guidance.

In short order the new bombs were en route to Afghanistan. There they would prove effective against al Qaeda and Taliban forces that had easily held the Soviet Army at bay in the mountain redoubts. The antipersonnel effects are particularly nasty, and the fine particles spread throughout the open areas, ignite, and suck all of the air out of the confined space.[4] The temperatures of fuel-air explosions can reach over 3,000 degrees Celsius, with a blast wave that travels at 10,000 feet per second.[5] Unlike in the movies, there is no outrunning the explosion.

There should be no doubt that thermobaric weapons will play decisive roles in future operations. Those who once believed they were safe from U.S. bombing in underground bunkers should expect to become prime targets. The importance of the high heat should not be underestimated. One of the biggest concerns in striking biological weapons facilities has been the potential for spreading the material during the explosion. Most BW agents would be thoroughly cooked if struck by a thermobaric device.

There are a family of bunker-busting bombs currently available and probably more under development. A penetrating cruise missile, the AGM-86D, has enhanced explosive capability. It also has a fuse that counts. When the missile hits a building, it penetrates the assigned number of floors and then detonates. Just because an adversary has a lot of building between himself and the roof does not prevent the bomb from withholding detonation and exploding at his level.[6]

Precision has been the name of the game for more than a decade. Those who watched the Nintendo aspect of Desert Storm remember the pictures of laser-guided bombs destroying buildings at will. The reality was that on many occasions bombs could not be dropped or the laser lock was lost due to cloud cover or battlefield haze. Since then the Joint Direct Attack Munition (JDAM) has joined the inventory. The first launch was made from a B-1B at Edwards Air Force Base, California, in February 1998.[7] The $18,000 JDAM has a tiny GPS receiver that can be programmed to direct the bomb to the target with accuracy of a few feet from up to fifteen miles away, all based on the signal from satellites.[8] First used in combat over Bosnia, JDAMs became a mainstay of the air campaign in Afghanistan.

In addition to new JDAMs, the Air Force has developed a guidance kit that can be affixed to conventional *dumb bombs*. These $14,000 conversion kits transform the free-fall bombs into precision weapons. In Kosovo they reported that more than 600 converted bombs were dropped, with 95 percent reliability.[9] The kits mean that these bombs can now be dropped in inclement weather.

Despite their remarkable accuracy, precision weapons have not been perfect. The most highly publicized accident was the bombing of the Chinese embassy in Belgrade, Yugoslavia. Several explanations emerged. Some said the bomb had gone off course. Another stated the maps were wrong and the building improperly targeted. Still others suggested it had not been an error but rather a message to Beijing. Enduring Freedom also had tragic accidents. In one, three U.S. Special Forces soldiers and five friendly Afghan fighters were killed when a B-52 dropped a 2,000-pound bomb about one hundred meters from their position near Kandahar.[10] The accident was attributed to a battery going dead in a *plugger*, the GPS receiver, as the B-52 was inbound. When the batteries were changed the coordinates automatically reset to the location of the combat controller, not the target he had computed. The problem was human, not technological. Earlier four civilians had been killed and eight wounded when a bomb hit a residential area near Kabul.[11]

Despite the few errors that have happened, JDAMs have held collateral casualties to a minimum. They will play a key role in all future conflicts. However, the most important issue will be the ability to identify the target.

As an adjunct to ordinance development, the Department of Defense is improving the navigational capability of guided munitions. Concerned that a sophisticated enemy might be able to jam the signals from the Global Positioning Systems (GPS), they plan to turn up the power so that bombs and missiles do not lose their targets and troops on the ground can be assured of accurately locating themselves.[12]

An innovative adjunct to the use of guidance modification has been proposed by Bud Kinsey of Raytheon. Called Big Gun, the concept would turn a traditional cargo aircraft into a weapons platform—one that could carry many times more munitions than current bombers.[13] Orbiting high over the battlefield, Big Gun would respond to calls for fire from ground personnel or UAVs that find targets. Each munition, such as the 105mm Multi-Role Armament and Ammunition System (MRAAS), is much smaller than larger bombs, and an aircraft could carry 1,500 rounds or more. There is no need for a launching mechanism, as gravity provides the necessary velocity. With a 2:1 glide ratio, from 40,000 feet the munition could hit targets in a sixteen-mile radius. Not a bunker buster, Big Gun would be effective against tanks and artillery and for suppressive fires against troops or surface-to-air missiles (SAMs). GPS-guide rounds can be delivered against targets that are located by troops or robots and the location transmitted to the aircraft via secure data link. Terminal homing seekers will ensure the round hits the designated target. Further, with rapid response time engagement can begin within minutes—faster than most conventional air or missile support.

An all-weather system, Big Gun would be ideal for supporting small special operations teams that might run into trouble during unexpected ground contact. The magazine is big enough to keep an enemy from overrunning the team until extraction can be accomplished. Similarly, the most vulnerable time for early entry forces is when they have only a few troops on the ground and have not established sufficient fire support from integral artillery. The long-endurance Big Gun could cover that operation for many hours at a time. It provides massive firepower without the need to deploy heavy guns.

Big Gun would require protection from long-range SAMs and does require an aircraft. Recognizing that strategic lift is a critical issue, the mission would have to compete for priority. Since the Big Gun system is aircraft-type independent, many planes could be used to provide small units heavy-precision firepower.

In the bunker-busting efforts, even nuclear weapons have been considered. In a significant departure from previous policy, the United States has indicated it will research nuclear weapons designed to eliminate deeply buried targets that might house weapons of mass destruction.[14] As good as the thermobaric weapons are, some of the deep targets will require more energy to destroy them. Given the precision delivery mechanisms currently available and the penetration that can be achieved, small nuclear weapons can be developed that take out the desired target without fear of excessive radiological fallout. It should be noted that the decision to research these weapons does not mean we will build or employ them.

The Air Force is not alone in improving their ability to strike from above. The Army has been working on the RAH-66 Comanche for several years. Plagued by management problems, the program has not moved as smoothly as desired. However, Comanche will soon add greatly to the Army's capability to penetrate hostile territory on both reconnaissance and attack missions.

The Comanche will have longer-range (1,200 nautical miles) and be faster (175 knots) than any of the existing military helicopters.[15] Its low observable characteristics make it ideal for evading air defenses. Though not as heavily armed as the AH-64D Apache, Comanche can both find targets and destroy them with its Hellfire missiles and 20mm cannon. Safe from most small-arms ground fire, the airframe is designed to tolerate hits from up to 23mm guns. The electronic countermeasures suite includes the Advanced Laser Warning Receiver as well as radar warning and jamming devices. It also has its own air-to-air defense capability when armed with Stinger, Starstreak, or Mistral missiles.

FROM NEPTUNE'S LAIR

A few years ago, the U.S. Navy and Marine Corps jointly published a pamphlet called *Forward . . . from the Sea* detailing the emerging and traditional responsibilities of the U.S. Navy.[16] The old guard argued that the Navy should be focused on keeping the sea-lanes open. Other senior officers outlined the need for more extensive operations in the littorals. Some called this a transformation from the *Blue Water Navy* to the *Brown Water Navy*. In reality, both missions are essential.

For decades the Navy's basic building block has been the Aircraft Carrier Battle Group, a mighty armada that provides the ability to project and apply power. Some military analysts believed the days of the Carrier Battle Group were numbered due to their vulnerability to long-range missile attacks. Of course the carriers were provided extensive protection with picket ships posted many miles away and air cover above and submarines below the surface.

It has been our experience that these Carrier Battle Groups have been a centerpiece for many recent U.S. military operations. The Navy, not the Air Force, flew the dominant number of missions in support of Enduring Freedom. The reason was because they could move their ships into position without need for basing rights or needlessly extending the flights of combat crews. True, the B-2s did make strikes in Afghanistan, but their missions were in excess of seventy hours for a single bomb load. Other fighter-bombers and C-130 gunships were later stationed in Kazakhstan, but that required weeks of negotiation and large payoffs.

The Navy was not alone at sea. With them were the versatile U.S. Marine Expeditionary Forces capable of both air and ground operations but fighting as a combined arms unit. Time after time in recent years they have been called upon to perform a variety of peace support and combat operations. From Africa they extracted civilians caught up in local rebellions. They covered the extraction of United Nations forces from Somalia and were among the first conventional units to fight in Afghanistan. The common factor was that no two missions were the same.

These operations prove that the Navy will continue to play a pivotal role in future combat missions. Geography is on their side. Nearly 70 percent of the population of the world live near major bodies of water, many of them in densely populated urban areas. Geopolitical realities suggest that we will be forced to operate from these waters as fickle allies provide and then withdraw support for any given operation. We simply cannot rely on the goodwill of other countries for basing, overflight, and ground operations rights. That means we must maintain our ability to attack from sea-based platforms. Unfortunately, like the other services, the Navy has been cut to the core. To keep a forward presence in potential trouble spots around the world, crews are constantly deployed, while ships and planes suffer from less than optimal maintenance. While personnel are more than willing to meet any exigent requirement, constant patrols without action quickly wear on them.

We are facing a paradox in our ability to maintain forward deployed forces. Funding is constrained, and we must put the money available to the best use. Therefore, we can build new foreign bases, expand existing ones, and pay off local governments so they accept our presence. Then we should hope that when the time comes to act they don't pull what the Saudis did and disallow missions from their soil.[17] Or we can put the funding into building new equipment, including high-tech ships that allow us to project power from the sea. The reality is that we will do some of both. The greater advantage would be in emphasizing research, development, and acquisition of new platforms and technologies that provide us the greatest degree of freedom to act. In this era of transient alliances we will be far better off developing a force that can act unilaterally when necessary.

Long gone are former Secretary of the Navy John Lehman's dreams of a 600-ship fleet, one that could dominate the seas at will. After the devastation of the Clinton administration the Navy is down to 314 ships and only about 385,000 personnel on active duty.[18] Rebuilding will take both time and money. In keeping with the notion of transformation, new technologies will be incorporated into the ships of the future as they prepare to fight the inevitable coastal battles.

On the drawing boards are several new classes of ships that will dramatically improve the Navy's capabilities to fight on the coasts. Among them are the Zumwalt-class land-attack destroyer (DDX), the Littoral Combat Ship (LCS), a new class of cruisers (CGX), and the Virginia-class submarine based on Seawolf technology but designed to get close to shore.[19]

The common characteristic of each of these new classes of fighting ships is that they can operate safely in the littoral areas and project power ashore. A few years ago, Vice Adm. Art Cebrowski, the president of the Naval War College, and Capt. Wayne Hughes suggested rebalancing the fleet. The basic concept was to build a number of smaller ships instead of relying on a few larger ones. These smaller ships could go in harm's way whereas the large, expensive ships were too valuable to risk.[20]

Streetfighter, they suggested, would be one such improvement. The cost would be less than 10 percent of a battleship, possibly $90 million each. Fast, the Streetfighter would travel at fifty to sixty knots, nearly twice the speed of the recently commissioned Arleigh Burke–class destroyers that topped out at thirty-two knots.[21] The range would be 4,000 miles, but more important, the draft would be less than ten feet, which would allow them to operate in very shallow waters.

To support special operations forces two Trident submarines will be converted from their traditional strategic missile role. An estimated cost of $3.34 billion will be involved in the complete transition from the Tridents basic load of 24 C-4 nuclear missiles to the capability of holding 154 Tomahawk cruise missiles and sixty-six Navy SEALs or Army Special Forces troops.

In support of the SEALs is a new Advanced SEAL Delivery Vehicle that can be satellited with the submarines. The ASDV is a great improvement over the earlier craft. First, the team is dry during transit. In the past, the SEALs were wet for the duration of the trip. Any diver knows that cold water will drain strength from even the strongest swimmers.

While fighting ships may seem romantic, a critical Navy mission will be increasing our strategic lift with Fast Sealift Ships backed by a huge amount of medium-speed roll-on roll-off capability. The fastest cargo ships in the world, with speeds up to thirty knots, the Fast Sealift Ships can transit the Atlantic in six days. However, all eight ships of that class can only move one mechanized division. Getting to the Persian Gulf takes eighteen days if the Suez Canal is accessible. While the Navy has been converting craft and building new Large Medium-Speed Roll-On/Roll-Off ships far more will be needed if mobility of large forces becomes necessary. With a shrinking U.S. Merchant Marine fleet, maintaining adequate sealift will be a major challenge.

While the price tag is high, the costs will not exceed what we would pay for international cooperation that cannot be counted on in a crunch. The American people would be staggered if politicians were uncharacteristically honest and would consolidate all costs associated with foreign intervention. Currently not counted are all of the side deals made to countries not directly involved in combat operations but whose acquiescence is desirable. To conduct Enduring Freedom many countries were rewarded with foreign assistance packages that are not attributed to the operational expenditures. As these do not come from the defense budget, they are conveniently left out of the totals. What is needed is to balance the entire sheet, present it to the people, and let them decide where the money should be spent. History has shown they will come down in favor of a stronger defense program rather than distributing our resources piecemeal.

ALWAYS THE GRUNTS

Desert Storm was indeed a turning point. After a six-month buildup phase, the allied forces struck out across the sands. While airpower had wreaked havoc, the Iraqi military was totally unprepared for the invasion that followed. For eight years in the 1980s they had held the Iranian armored forces in check. Certainly, they thought, they could hold their own against American armor. Capable of cross-country speeds up to forty-two miles per hour, while firing with devastating accuracy on the move, the M1-A2 Abrams tanks ruled the day. Iraq may have lost 3,700 of its Soviet-made main battle tanks, many of them to these seventy-ton behemoths.[22] Even now there are upgrades for this monster. The highly classified armor can withstand direct hits from most other tanks. The 120mm gun provides first-shot kill capability in excess of two miles.

In May 2002 the Army canceled the 155-self-propelled-howitzer Crusader program that had already demonstrated greatly improved artillery capability. The integral midwall cooled tube permitted firing rates of ten rounds per minute. With range to forty kilometers, Crusader offered the unique multiround simultaneous impact (MRSI) in which eight rounds from the same gun would strike the target at one time.[23] After a rather public acrimonious battle between Army proponents and the Secretary of Defense, the program died literally of its own weight.

The new army, nominally called the Objective Force, wants things light. The Abrams and Crusader are just not deployable. The heavy-lift C-17 Globemaster III can only carry one system at a time. It took a month to move a single division by sealift from the United States to Saudi Arabia during the Gulf War. In today's rapidly changing environment, to be effective you must get firepower to the operational area, and get it there fast.

Mindful of the realities of combat in the near term, the Army supported two innovative megaprograms to provide a major leap ahead in fighting capability. One is the Future Combat System (FCS), basically a new brigade but with design starting from a clean sheet of paper. The second is the Objective Force Warrior (OFW), a system of systems that would equip the soldiers who would man the FCS.[24]

The requirements for the FCS are ambitious. It is to be smaller, lighter, sustainable, and more lethal than its current counterparts. Sensors, advanced communications, and agility are key to survivability. There is to be nothing over twenty tons in weight, and ten tons is preferable. Also, it must fit on a C-130 aircraft. The total FCS weight on departure will be less than 6,000 tons, well under half that of existing brigades. Further, with about one-half the personnel, it is still to be deployed in ninety-two hours.

Once on the ground the FCS must be able to fight for a minimum of seventy-two hours without resupply. The vehicles will be very fast. They are required to travel 125 kilometers per hour on-road and 90 kilometers per hour off-road (that's 78 and 56 miles per hour, respectively) and they are expected to be able to cover more than 450 miles in twelve hours. Once in a fight, the FCS will be able to engage an enemy at long range, as much as seventy kilometers away. Emphasis is on the ability to strike targets beyond the line of sight. This is accomplished by placing remote sensors on both ground and airborne platforms that will provide early warning of an advancing threat. The sensor information is passed to the shooters via a very sophisticated command and control system known as the C4ISR architecture.

Unmanned systems will be essential. The FCS plans for a family of un-manned aerial vehicles that provide sensing and limited attack capability. Ground unmanned vehicles will also vary in size. Some will have guns and be able to fight. Some will be sensors that can be deployed in dangerous situations. Others will transport equipment that is needed after periods of combat. There also has been consideration of a robotic medical evacuation vehicle, one that can drive over a wounded soldier caught in the open, scoop him up, and bring him to safety.

An innovative concept surely to be included in FCS is NetFires. This was developed and called *missiles in a box,* as it was a self-contained package of sixteen tubes that could be dropped into any location. Fifteen of the tubes had precision attack missiles, while the last tube held the fire control system. Weighing 2,500 pounds, the entire system could be delivered by a C-130, Ch-47D, or on the ground by an HMMVW.

The NetFires missiles could be either fired on request of soldiers observing a target or sent out in a loiter attack mode. The Loiter Attack Munition (LAM) would be launched vertically; wings deploy, and then it glides out forty kilometers. Once in the objective area, the LAM would search for targets as it stayed aloft for up to an hour. The LAM would be used when the remote sensors indicated an enemy armored unit was approaching. The approximate rate of advance would indicate where the LAM should begin the search pattern. The automatic target recognition sensors would find the specific tank and initiate the attack.

The FCS is designed to accomplish the Army's stated goal, "*See First, Understand First, Act First, and Finish Decisively.*" It cannot be viewed as a single piece of equipment or even a system. To be successful the FCS will be a fully integrated system of systems highly reliant on speed, agility, and the ability to remotely kill enemy armor.

In the near term the Stryker family of vehicles promises to meet the mobility and lethality goals. Named after two soldiers named Stryker, both of whom won the Congressional Medal of Honor, these vehicles will provide an interim step. At nineteen tons, Stryker barely meets the weight limitation for C-130 air mobility. It comes in a variety of configurations so that the same basic vehicle can accomplish many missions. This is key to maintenance and overall sustainability of the unit. Options include a mobile gun system, infantry carrier, antitank platform, fire support vehicle, engineer squad vehicle, and recon vehicle, and the Stryker can be used for medical evacuation.

The Stryker has eight wheels, thus ending temporarily the wheels-versus-tracks competition. Among the weapons choices are an M68A1 105mm cannon, TOW 2B missiles, several mortars, and a variety of machine guns or grenade launchers. It can travel sixty-two miles per hour and climb steps up to twenty-three inches high. The infantry carrier version is said to be able to deliver troops to the battle while they are in reasonably good condition.[25]

Billed as soldier-centric, the Objective Force Warrior (OFW) is the person who will engage our enemies. The notion of soldier-centricity suggests that the requirements of the individual fighter take precedence over technology. With the OFW concept the Army is equipping the man, not manning equipment.* That is an important distinction, one too often forgotten in earlier weapons system development. In the end, this soldier of the future will be like no other before him and will have more resemblance to Robert Heinlein's *Starship Trooper* than the snake eaters I fought with in Vietnam.

The goal for the OFW is to get individual equipment weight down to under forty-five pounds from his skin out.[26] To put that in perspective, about fifteen years ago I served on an Army panel to *lighten the soldier's load*. At that time the *lightest* individual load for an infantryman was 128 pounds. The loads for machine gunners and a few others were considerably heavier. Real combat loads, laden with extra ammo and water, easily could run 100 pounds. Technology had not been the soldier's friend. Every time technology

---

*Editorial note: I am using the male pronoun with the OFW as I believe we will keep our restrictions on a male-only combat arms positions (armor, infantry, and artillery). It seems unlikely that this will change in the next twenty years. I could be wrong, and the Army has made other mistakes in the name of political correctness.

made an item lighter, we found an excuse to add other items, nearly always ending up with a net gain in weight.

Key to the OFW is situational awareness. That's military jargon for knowing what's going on around you. Integrated into the FCS, the soldier will have helmet-mounted displays and audio input that provide all of the information he needs. This is seen on the two stereoscopic pictures inside the visor that protects his eyes from fragments and enemy lasers. The helmet will include miniature night vision sensors and have built-in antennae.

Using the satellite GPS system, the OFW will know where he is and the location of other members of his squad and be able to distinguish between friend and foe (IFF), all marked on his visor. More complex will be identification of neutral parties, or noncombatants. The IFF issue is critical. In Desert Storm it was estimated that about 25 percent of coalition casualties were from friendly fire. Operation Iraqi Freedom proved equally fratricidal.

The estimated cost for equipping each soldier is about thirty thousand dollars—not cheap. Much of that is in sophisticated electronics and the computers he will be wearing. There is a personal data system on his left arm that plots the information and maps his position. The OFW has a totally unique communications system. The Joint Tactical Radio System (JTRS) is netted so that individuals at all levels of command can communicate. This advance will eliminate the endemic problems that plagued previous soldiers when one unit could not talk to another.

Power to run all the electronics is critical. The OFW uses nickel metal hydride batteries that have been called *cell phone batteries on steroids*. Photovoltaic solar cell patches in the outer uniform will collect power in daylight. In addition, using the striking of his heel each soldier will generate power with each step he takes. He should be able to fight a minimum of seventy-two hours before replacing or recharging his power supply. This entire electronic suit is expected to weigh about four pounds.

The body armor, capable of stopping small-arms fire, is suspended about two and a half inches from the skin. This better distributes weight and allows for deformation on the inside of the armor with serious injury if bullets hit the soldier. To maintain his body temperature, even in extreme climates, the suit has microclimatic conditioning that can heat or cool him as necessary.

The system will also monitor physiological data from the soldier. These include blood pressure, heart rate, and core body temperature. The system can also evaluate his caloric intake and compare that with food consumption. This provides data on how much physical energy the soldier has.

If a soldier is wounded, the information will be passed to his comrades. The extent of physical injury will be noted, including death. There will not be any guessing as to whether rescue should be attempted. In addition, the system can provide automated medical triage and combat casualty care. For instance, if a severe wound or traumatic amputation to a limb is determined, then the system applies an embedded tourniquet to stop the bleeding. In addition, the suit provides protections from chemical and biological attacks. A slight overpressure in the helmet protects the soldiers from breathing in toxic materials.

To enhance survivability in combat, the OFW will have adaptive camouflage. Just as the creature in the science fiction movie *Predator* gained invisibility by becoming transparent, the uniform will sense the surroundings and change color, allowing the soldier to blend into his background. This will be accomplished automatically and require no attention from the fighter.

In past wars the infantryman had to carry the weapons with which he was going to fight. Defeating enemy armor or bunkers often required systems bigger than could be carried by the individual. The OFW still has a personal gun, but through JTRS integration with the FCS he has the ability to direct heavy firepower that is immediately available to him. No longer will he be required to go through a liaison officer and hope that the artillery will shoot for him.

To lighten the carried loads, the OFW comes with a robotic mule. This system will follow the infantry squad and bring the bulky equipment. It is targeted to weigh 500 pounds and run on a hybrid electric and JP8 fuel. That engine is loud and consideration is being given to a stealthy all-electric version. The mule can carry extra ammunition, food, batteries, and sleeping gear. In addition, it will be able to generate water, which is both heavy and critical to the fighting man.

On the horizon is an exoskeleton system that can enhance human performance. The objective is to provide superhuman strength to each soldier. With powered actuators, the exoskeleton allows people to move faster and lift more with less energy. The actuators are the robotic equivalent of muscles. The advantages to such a system are enormous. The soldier can carry larger weapons, wear more body armor, and leap from places previously unavailable. In urban areas, an exoskeleton-equipped soldier can get up stairs more quickly.[27]

Power is again the long pole in the tent. Running the actuators takes a lot of it. Fuel cells will be part of the answer as they continue to improve. When soldiers were first briefed on the Pitman concept over ten years ago, it was clear that no soldier wanted to be tethered to a power source. The technical hurdle at that time was power. That remains true today, but we are getting closer and DARPA is working on it.*

THE TRENDS

No single book can adequately cover all of the evolving weapons systems of the future. However, certain trends are discernible. These include significant emphasis on unmanned platforms, use of nanotechnology to make things very small, information warfare against nonmilitary targets, and asymmetric applications of weapons of mass destruction. From an American perspective, we want things precise but remote. Our adversaries want to engage us in ways that do not permit us to bring our full military might against them. Accommodating all potential threats is impossible. Thus defense requires a unique combination of technology and innovative soldiers.

GETTING PEOPLE OUT OF HARM'S WAY

In recent years the American public has become accustomed to the use of robots for a variety of tasks. While they are employed extensively in heavy manufacturing, such tools are rarely seen. However, during scientific exploration in harsh environments they have played a key role. Most people remember the small vehicle that became the first system to navigate on the surface of Mars and send back panoramic photos of the Red Planet. Undersea unmanned vehicles have been used to photograph the ocean's floor at depths not yet attainable by manned submarines. There they located heat vents and exotic life-forms previously unknown. In rescue operations the news has filmed miniature robots searching through rubble in places too small for humans to crawl. In the Great Pyramid in Giza a robot was sent to explore tiny passages in search of hidden vaults.

All of these applications can be directly transferred to the military. In fact, many of them were developed under military research contracts. Familiarity has changed perceptions about the use of robots. Not long ago Congress forbade even research into a robotic system that had a kill mech-

---

*John Alexander and Jeff Moore, Pitman was a concept for an exoskeleton system developed by Jeff Moore at Los Alamos National Laboratory. With Gen. Paul Gorman we briefed Pitman to the commanding general of the Infantry School and Center. We were a decade too early.

anism. Yet during Enduring Freedom the armed Predator demonstrated the effectiveness of remotely piloted airborne craft that could find a target and shoot on demand.

There are many combat operations that can best be handled in an unmanned fashion. Robots are the talk of all military forces. Like UAVs, their jobs, known as 3D missions, can be connoted as *dirty, dull,* or *dangerous.* In an effort to conserve out fighting strength, every effort is being made to minimize exposure of troops to high-risk situations. All services are aggressively pursuing application for robots in their mission areas.

Robots are ideal for going into *dirty* contaminated areas in which chemical or biological agents are suspected. They can safely monitor the surroundings and report back to the command post using an array of sensors. Unlike people, robots do not get tired or become complacent. Therefore, they make excellent sentries for *dull* tasks and observe areas for long periods of time. Their main constraint is providing power. And robots can be sent into hostile areas that previously had to be explored and cleared by humans. Based on the probability of encountering a threat, they can cross *dangerous* open areas, be exposed to snipers, or enter rooms before troops arrive.

Size is important. Robotic systems can be quite small and yet accomplish their mission. For both underwater and aerial unmanned systems, removing a human means that life-support systems and redundancy for survivability are not necessary. This can reduce costs dramatically and robotic systems can be flown affordably in dangerous situations. This is a concept that the press seems to have missed. For example, in Bosnia and again in Afghanistan every time a drone was lost, the media played it as though a real aircraft had been shot down. The fact of the matter is, we will use unmanned aircraft at close quarters so that we don't lose pilots. Therefore, these aircraft can be assigned very risky missions without the possibility of a pilot being captured or killed. In short, if they get shot down, that is part of their job and may even be a success story.

Acceptance of the importance of robotic systems has not been easy. A decade ago when I recommended use of unmanned aircraft when lecturing to the U.S. Air Force Command and Staff College, I was booed by some of the pilots in attendance. The fighter pilots did not want to lose their preeminence. It was not until the Air Force Science Board made strong recommendations supporting the use of Unmanned Combat Aerial Vehicles (UCAV) that serious efforts got under way.

At first UAVs were perceived as an extension of reconnaissance operations. That was obvious. The additional missions, including attacking targets and providing logistical support, were much more difficult to sell in the Air Force. Under development currently are several new UCAVs. The X-45A being tested at Edwards Air Force Base is one example of a UCAV that will fight. Designed to carry 3,000-pound bombs, the X-45A is fairly large and has a wingspan of thirty-four feet and a unique Y-shaped tailless body.[28] It should be noted that theses UCAVs are not throwaway items. The estimated production cost is between $10 and $15 million each. However, that is about one-third the anticipated cost of the next-generation strike fighter.

The Army was very willing to explore unmanned aircraft. They desperately wanted to see over the next hill and to do it with assets that belonged to them. Therefore, they began to develop a whole family of Tactical UAVs.

Despite the grandiose promises, not everything was rosy. Some of the early attempts at developing an unmanned fleet for reconnaissance failed, and at high cost. There were many crashes and unsuccessful recoveries in the learning process. The difficulty in maintaining communication with the bird had to be overcome while payloads were reduced in size. Now the new fleet is beginning to fly and it will provide future commanders great opportunities.

Already fielded and deployed to Kosovo for tactical operations was the Hunter. Weighing 1,600 pounds, this is a large UAV with a twenty-nine-foot wingspan. It can orbit for as long as twelve hours and provide pictures of the terrain and enemy situation. In tests, a heavily armed Apache helicopter was able to remotely control this UAV, making them a very effective hunter-killer team.[29]

Among the new leaders is the Shadow 200, built by AAI Corporation. The Shadow has successfully participated in several operational tests. Fairly large, it has a wingspan that exceeds twelve feet and a sixty-pound payload. Flying up to 15,000 feet, the Shadow can stay aloft for up to five hours. There are a number of sensors and designators available for the Shadow. For intelligence gathering it may have synthetic aperture radar, moving target indicators, and electro optic infrared sensors. Once located, targets can be illuminated for strikes by laser-guided munitions. The Shadow will belong to maneuver commanders, providing them intelligence never before available in direct response to their needs.

While neither Hunter nor Shadow has been equipped with weapons, there are plans to add them. These UAVs could carry BAT, Hellfire, or Stinger missiles. Once this is accomplished a pressing issue will be establishing the rules of engagement under which the missiles can be fired. At present the ROE are overly restrictive. They require that EO sensors provide pictures with details as one would get through binoculars before soldiers pull the trigger. Unfortunately, SAR imagery does provide the clarity necessary for the decision to shoot. Commanders must learn how to interpret the SAR imagery and allow fires based on that data.

Other UAVs are being researched. Devil Ray, built by General Dynamics, is a smaller craft with only a seven-foot spread and thirty-five-pound payload. Northrop Grumman's Fire Scout is a variant that allows vertical takeoff and thus does not need the runway required by traditional UAVs. *Fire Scout* is being developed for the U.S. Navy and Marine Corps in order to provide precision targeting up to 110 nautical miles from the launch site. This can be any aviation-capable warship. With rotor diameter over twenty-seven feet, the Fire Scout will be able to fly to 20,000 feet with six hours' endurance.

Much smaller in scale are the micro-UAVs. By the DARPA definition, a micro-UAV is a craft of fifteen centimeters or about six inches or less in any aspect, height, width, or length. These can fit in your hand and would be ideal for troops who just need to see what is going on around a corner, behind the building, or in the trees ahead of them. It has been predicted that micro-UAVs will eventually get down to the size of a hummingbird. At the moment, micro air vehicles of six to nine inches have formally made it into the Advanced Concept Technology Demonstration (ACTD) list, signifying that they will likely be available in the near future.[30]

A few years ago the Navy developed *Sender*. It weighs about ten pounds and has a four-foot wingspan, well above a micro-UAV. Still, *Sender* has a range of nearly 100 miles. The intent is to bring the scale down one order of magnitude, fly up to sixty minutes, and have a range of ten kilometers. For small units and special operations forces micro-UAVs will be invaluable. As sensors get lighter and power sources improve, the micro-UAV will become a mainstay of reconnaissance for the frontline troops.[31]

Cost is an issue with many of our UAVs. There is a current effort to reduce the cost and provide a craft that can be dropped from a larger airplane and perform typical UAV missions. Previously it was known as the Loitering Electronic Warfare Killer. On the ACTD list it is called the "Expendable Unmanned Air Vehicle." This functions much like the comic-book Transformers. Stowed, it looks like a missile. However, once it is dropped wings swing into place and an engine kicks in, allowing the UAV to travel up to 1,000 miles or loiter for eight hours. The 200-pound payload can be configured with a combination of sensors and munitions. It is designed to be released from fighter-bombers, B-52s, helicopters, or a C-130 aircraft. With a target cost of $40,000 it is considered expendable.

Many other countries are actively engaged in development of UAVs. Some of the Army's early craft were bought from Israel. Currently several European countries are working on Euro Hawk, Eagle, Tailfun, Hermes, and Ranger, to name a few. Clearly UAVs will be a common system in all services from technologically developed countries. We can be sure other countries will buy off-the-shelf models as they become available.

There are drawbacks to the use of UAVs. Most have weather dependencies for both their flight characteristics and sensor operations. The lighter the aircraft, the more sensitive it is to wind. Obviously, the micro-UAVs will have trouble in any significant breeze. The larger Tactical UAVs have cross-wind takeoff limitations, and precipitation inhibits most of their sensors. Still, they offer a major leap forward for ground reconnaissance.

ROBOTS ALSO SWIM

For underwater operations the U.S. Navy has developed their Unmanned Undersea Vehicle (UUV) Master Plan.[32] As with their airborne counterparts, there are many missions UUVs may perform, including reconnaissance in dangerous, possibly mined waters, search and survey, serving as a communications aid, and tracking submarines.

Mine detection is one of the main missions. The first operational UUV will be the submarine-launched Long Term Mine Reconnaissance System, and that is scheduled for 2003. Underwater survey and bottom mapping in support of amphibious landings will take on increased importance as the Navy chooses to operate more frequently in the littorals. Lazarus and Seahorse, both fairly large autonomous unmanned vehicles, will be exploring those capabilities. They use acoustic communications for image transmission and position information.[33] Both UUVs are well in excess of twenty feet in length and about three feet in diameter. Considerable effort will go into reducing the size of UUVs as the missions increase.

Not all unmanned naval craft will be subsurface. During Desert Storm, Sandia National Laboratory developed a unique remote-controlled system called Surfwedge in six weeks. Despite the SEAL reconnaissance efforts, it was known that an amphibious invasion would likely encounter shallow-water mines. To support the U.S. Marines, Sandia equipped six old landing craft with kits for remote control. The intent of the operation was not sophisticated. These landing craft were sacrificial and would lead the attack. Their mission was to detonate the mines and sink. Although the amphibious operations were planned, they were not executed in support of the ground attack.

Current UUV considerations are more complex. The Spartan is a high-speed surface vessel being developed for antisubmarine warfare in the littorals as well as countermine and reconnaissance operations. With speeds varying from twenty-eight to fifty knots, Spartan can function up to forty-eight hours and at ranges to 1,000 nautical miles. It can carry payloads between 2,600 and 5,000 pounds, based on configuration. It will be equipped with various sensors and weapons depending on the assigned mission.[34] Spartan also made the ACTD list.

On a much smaller scale is a surface UUV called Roboski. It is just what the name implies, a robotically maneuvered Jet Ski. Available from Robotek Engineering, the Roboski was briefed to the Naval Science Board study on non-lethal weapons. In the wake of the USS *Cole* attack, there was a need to be able to inspect or interrogate approaching craft. As we now know, the small boat that attacked the *Cole* was packed with explosives. A Roboski could be deployed with chemical sensors that sniff downwind and determine if high amounts of nitrogen are detected, indicating a craft is carrying a bomb. Roboski offers an immediate solution at low cost and could easily be deployed from a larger ship.

UNMANNED GROUND VEHICLES

It was during Enduring Freedom that combat robots passed through their baptism of fire. Soldiers of the 82d Airborne Division used a remote-controlled vehicle named Hermes, after the god of innovation and cunning, to clear caves near Khost, Afghanistan. This was not a task that was eagerly awaited. It was from the eastern mountains that the Taliban had held off the Russian army for years. The caves are deep, winding, and treacherous. With years to prepare, there was concern that survivors might be hiding out or, worse, had booby-trapped the area before they departed. Older members of the U.S. Army remembered too well the stories of the infamous tunnel rats that entered the underground labyrinth constructed near Cu Chi, Vietnam. A tunnel-clearing mission is always dicey.

But now the troops had a forty-two-pound friend. About a foot high and three feet long, Hermes could crawl along the underground passages and search out the traps. Traveling on light treads, the robot responded to the commands of the operator waiting above. Two cameras provided imagery and allowed the robot to be steered around boulders and over small obstacles. That day, no dangers were discovered.

For less than $40,000 Hermes was built from commercial off-the-shelf parts. It can be fitted with up to twelve cameras or an array of sensors. A similar police robot carries a twelve-gauge shotgun, while military models may incorporate grenades for defense. Given the success of this first combat mission, it is highly likely that robots will be integrated into the force very quickly.[35]

For more than a decade the DoD Joint Robotics Program has been developing and testing various kinds of teleoperated and autonomous vehicles. Most appropriately, their motto is "Out Front in Harm's Way." All services and several foreign countries are participating in this program headquartered at Redstone Arsenal near Huntsville, Alabama. They explore robots for many different kinds of missions. There are the mundane logistics functions, moving heavy items and loading ammunition. Explosive ordinance disposal (EOD) units have long worked on developing and employing robots to interrogate suspicious packages or remove known bombs. In conjunction with EOD activities has been the introduction of mine-clearing machines. Placed on a tanklike chassis, these robots are very heavy and can sustain the blast from mines. They have been successfully deployed in Bosnia, where they found numerous mines without any human casualties.[36]

Other missions include reconnaissance and area security. Robots' use in sniffing out chemical agents and explosives will save many lives. But now there is greater emphasis on creating a family of robots that will actively participate in ground combat.

Fire Ant is one such innovative program at Sandia. Armored vehicles move easily cross-country and do not always go where they are expected. Therefore, static minefields have value only in keeping tanks out of specific areas. Fire Ant is a mobile mine that can detect moving armored vehicles and position itself to attack them. Placed in a position of tactical advantage, Fire Ant's sensors watch for the moving target. Once the target is located, the robot fires an explosively forged projectile at a weak point on the target. The blast from the weapon also destroys the Fire Ant. However, this is a very cost-effective and safe method for engaging enemy tanks.[37]

Another example of Sandia's innovation is their Surveillance and Reconnaissance Ground Equipment (SARGE). Built for the joint program office, SARGE is mounted on a Yamaha Breeze all-terrain vehicle for cross-country mobility. The sensors currently include day/night and thermal imaging.

SARGE has a high-power zoom camera that can resolve a man walking at a kilometer. The sensor packages can be altered to fit the mission.

Sandia has also developed a fleet of robots for security purposes. About one foot in length, these robots have both sensors and communications equipment. They are smart enough to respond as a group for area coverage. Based on the communications each receives, they can autonomously adjust position so as to ensure continuous surveillance of their assigned area.

There are many companies now engaged in manufacturing robots and developing new ones. Increasingly robots are being introduced into field exercises. Troops are gaining experience in how they may be best employed. Clearly they can go into areas inaccessible to, or too dangerous for, humans. Of course there are countermeasures. For small devices, a heavy blanket can stop movement and block optical sensors. However, robots will be invaluable for clearing buildings, tunnels, and caves. The biggest question will be who operates them. If a soldier is controlling a robot there are other tasks he is not performing. Still, robots are the way of the future and they will proliferate quickly.

THINGS SMALL AND SMART

Size does matter. When it comes to war bigger is not always better. One of the major revolutions in warfare will be adaptation of nanotechnology, also called molecular engineering. These items will be built by manipulation of materials at the atomic level and will include manufacturing of electrical circuits and mechanical devices. Some experts believe that weaponization of nanoscale robots will have greater impact on warfare than nuclear weapons. Adm. David Jeremiah, former vice chairman of the Joint Chiefs of Staff, several years ago stated that "military applications of molecular engineering have even greater potential than nuclear weapons to radically change the balance of power."[38]

The implications go far beyond war. It is believed by some scientists that microscopic robots, or nanobots, as some call them, may dramatically alter the future of civilization. The science and engineering will create systems that may be used either positively or for harm. Nanobots will be so small they could be injected into the body and targeted against blood clots, plaque, or other harmful substances. They would seek out the offending agent and begin to disassemble it, thereby providing lifesaving assistance.

However, nanobots can be sent to disassemble almost anything. They can be employed on weapons systems whereby they are sent out to inspect, mend, or destroy the targeted substance.[39] Since they will make microscopic sensors, these nanobots may become a first line of defense against biological attacks by terrorists. At one-hundredth the width of a human hair, nanofibers can be produced than reinforce polymers, making materials both stronger and lighter. It is estimated that nanofiber helmets would be 40 to 60 percent lighter than the standard model today.[40]

In reviewing technology trends for the National Intelligence Council, the RAND Corporation identified three that seem to be critical. One was genetically modified food that will enhance production while minimizing attacks by pests. On that issue they noted problems associated with public acceptance. While the media will be addressed later in this book, the insistence on calling them *Frankenfoods*, as some reporters do, represents ergofusion and does a disservice to people who need these products. The other two technologies identified relate directly to this portion of our discussion. They are smart materials and nanotechnology.[41] It is these two that will impact security.

Both a major advantage and disadvantage to nanobots is their ability to self-replicate. That is, once constructed these molecular machines will be able to make more nanobots capable of making even more nanobots. Robert Merkle, one of the leading researchers in the field, notes that while the research and development costs are high, self-replication will bring manufacturing costs down to an acceptable level.[42] William Tolles, recently retired from the Naval Research Laboratory, has written extensively on the importance of self-assembled materials. He states that they may be the building blocks of the twenty-first century just as alloys, plastics, and semiconductors were in the twentieth century.[43]

The concern of other people is that self-replication will extend beyond control of the designers. Tolles also indicated that use of nanotechnology-based weapons systems against an adversary without that technology could "be stunningly lopsided in their outcome."[44] Countries with developed nanotechnology infrastructure can establish defenses against them and employ the technology against other weapons of mass destruction. For the next few decades the technology base will limit the participants.

Calling nanobot weapons a negative use of the technology, Andrew Chen has already begun the counterspin. He correctly indicated that nanoweapons could "miniaturize guns, explosives and electrical components of missiles" and that "disassemblers to attack physical structures or even biological organism at a molecular level" could be developed.[45] However, his argument that the technology is inherently bad fails, just as it has with non-lethal weapons. Further, it is unlikely that such research will be banned, nor would adversaries adhere to a ban.

NASA-developed synthesized carbon nanotubes are one of the items that will herald in an age of nanotechnology. They are very light, flexible, and extremely strong and have no physical limitation on length. About one hundred times the strength of steel, carbon nanotubes have about one-sixth the weight. They can be used to store energy, build structures, and even deliver drugs to soldiers.[46]

Only slightly larger war machines than nanobots are the micro-electromechanical systems (MEMS). Led by DARPA, considerable effort has gone into making these tiny devices that will become as prevalent as today's microprocessors. MEMS can place an entire system on a single chip including power sources, transceivers, sensors, and a microprocessor. The intent is to get the entire system down to the size of a grain of sand.

An early function for MEMS will be sensing. They will be small enough that they can be scattered about a battlefield or any other designated area and report when a selected material is present. They have been described as smart dust. Floating in the air, these minuscule sensors on silicon will be light enough to stay aloft for hours.[47]

There are many military applications for MEMS. Very rugged devices, they will make smart weapons smarter through enhanced precision. MEMS will provide navigational aids for weapons and people and improve communications systems. They are integral to the X-45A UAV, which incorporates 10,000 MEMS sensors, microprocessors, microelectronic devices, and microactuators in place of control surfaces. Microjects will make rounds go farther.[48] Via MEMS microclimatic conditioning, both heating and cooling can be achieved in the uniform of the future warrior.

The United States and Europe are not alone in exploring the value of micro- and nanoscale weapons system. In *Chinese Views of Future Warfare*, a book edited by Michael Pillsbury, the importance of these technologies is addressed. Maj. Gen. Sun Bailin detailed the use of ant robots, blood vessel submarines, and nanosatellites. He notes that this technological revolution was "transforming man's relationship with nature" and that "we have progressed from the material millimeter-micron stage to the molecular-atomic nanometer stage."

Just as making weapons systems smaller is a significant trend, new classes of smart materials will also change the makeup of these systems. While many of these materials are strong and flexible, they also have the ability to sense faults and autonomously begin a self-healing process. In traditional polymers, once cracks appear the items are weakened. The new materials begin the healing process immediately. Soldiers and their weapons systems are constantly exposed to dangers that might reduce effectiveness. These self-healing materials have great value in keeping soldiers available for combat even after injury has occurred.

Another application may be shape-memory polymers. These are plastics with memory that will regain their original configuration when physical conditions change. A small plastic ball could be dropped or placed in a location where a vehicle is needed. When the sun warms it, the shape-shifter would transform itself into the desired object and continue on its mission.

While still a ways from fruition, the basic research that will allow integration of smart materials into weapons systems is well under way. Like things small, things smart will play quite a role in the transformation of conflict.

WEAPONS OF MASS DESTRUCTION

Usually defined as chemical, biological, and nuclear weapons, weapons of mass destruction are given special priority by politicians due to their ability to kill large numbers of people. In general, our thinking about such concepts as mutually assured destruction (MAD) has changed. The threat is no longer that we might exterminate life on the planet. Rather, we are concerned about extensive damage and loss of life should any weapons of mass destruction be used. They are of concern not so much because the United States or Europe might cease to exist, but use of such weapons would precipitate wider conflict.

Many volumes have been written on these technologies both individually and collectively. There is not much to be added in addressing them again with brevity. They are alternative systems that must be factored into any decision to engage in armed conflict. A trend has already emerged for terrorists and others to attempt to obtain weapons of mass destruction for their arsenals. This proliferation does not bode well for future stability. Any traditional adversary would be ill advised if they ever attempted to use such a weapon against us. Should a major chemical, biological, or dirty nuclear attack be made against the United States it is likely that public indignation and anger, so far relatively muted, would not only support massive nuclear retaliation; they would also demand it.

PART II

THE REAL WORLD

"STAR TREK MEETS THE History Channel" might be the log line for a new television series based on future conflicts. Unique juxtapositions of high-tech and ancient warfare techniques will become a norm. During Enduring Freedom, American forces dropped precision-guided bombs to compliment horse-mounted cavalry charges accompanied by indigenous garbed Special Forces soldiers. If past is prologue, the future will continue to bring unanticipated anachronisms limited only by the ingenuity of both weapons developers and innovative soldiers.

The following chapters present a series of hypothetical scenarios that allow the reader to experience possible futures. All places discussed are real. The geopolitical situations have been only slightly extrapolated from current events. None of the future weapons systems described is unobtainable and most now exist in some stage of development or fielding.

The locations were chosen for the following six chapters as representative of the world that could be. They are important as they provide insight into future headlines and will make the reader better equipped to understand the actions of both political leaders and our military. These are not the only situations that could occur. In the asymmetric conflicts that are sure to come, the methods of attack are virtually unlimited.

The situation described in Egypt represents the problems the government has met in countering internal elements that wish to destabilize the country through disruption of its lucrative tourism industry. That conflict continues and could erupt at any moment. The ship scenario is based on the devastating terrorist attack on the USS *Cole* as it was docked in Yemen. Vulnerabilities still exist even with enhanced protection. The story of drug interdiction in the Peruvian Amazon looks at the actions taken in a futile attempt to stop importation of drugs from that remote region. For unfounded ethical reasons, these operations will fill gaps between other open hostilities. Finally, the hostage situation depicted in Nepal could happen in any of many developing countries where small groups of relatively rich tourists collide with internal political turmoil and endemic poverty.

The remaining two chapters address major war. While the locations have been artificially imposed, the concepts are based on actual preparation for combat. They raise serious issues that transcend events on the battlefield. There is no doubt that American forces can defeat any specified military threat in a conventional fight. Our special operating forces have done a magnificent job in preparing for these conflicts, but the services have not kept pace. Sooner or later expanding involvement in international crises will meet the finite limits of resources available. "The War We Will Fight" addresses some of those issues. It also provides an asymmetric option that may be the only solution to reestablishing stability in part of the world. Most likely, our failure to respond to these threats now will be paid for in American blood, just as we have learned in mobilizations past.

CHAPTER FOUR

THE NILE

DRESSED IN AN OLIVE *green galabiya, the traditional garb of Egyptian men, the captain of the* Oberoi Imperial *inhaled deeply on his Belmont cigarette. In his fifties, the heavily mustached man had been navigating the river since he was a small boy and learned the trade from his father. "Allah be praised." With a toot on the horn and both hands raised, palms facing outward, the captain acknowledged the passing barge, which chugged slowly southward against the formidable current of the Nile. In April the weather is quite moderate. In the old days the European and American tourists would fill the many cruise ships that carried them on journeys to fabled lands both above and below the mighty Aswan Dam. Though these ships were built to handle more than one hundred guests in air-conditioned comfort, they were now lucky to get more than twenty for any given trip. And this ship was one of the few still operating.*

Unrest in Egypt is nothing new. The infamous massacre of tourists at Luxor's Hatshepsut Temple on November 18, 1997, brought international attention. The extensive publicity dramatically impacted tourism, a great financial loss to the poverty-stricken country of Egypt. Though saddening, it was not the sixty tourists of various nationalities and police gunned down near the temple who constituted the deepest loss. Nearly 700,000 Egyptians worked the tour industry and another 7 million indirectly benefited from it. With 18 percent of the country's economy based on tourism, the outlawed terrorist group al-Gamaa al Islamiya struck deep at its real objective, to bring more instability to this fledgling democracy. In response many members of the group were rounded up and the guilty perpetrators tried and executed.[1]

Taking the threat of terrorism seriously, Egypt took swift action in providing protection to the magnificent sites. Guards were everywhere—at the entrance areas, on the buses, and inside the monuments themselves. Standing on the Giza plateau near the three great pyramids one could scan the horizon with binoculars and see soldiers riding camels establishing a broad perimeter of security. Armed with old Soviet AK-47 Kalashnikov rifles and radios, they spent hours patrolling the burning sand, establishing a wide perimeter just to make the tourists safe. With great operoseness the country that lives off its dead began to revive the trade, initially off over 90 percent. After four years the tourist industry had approached 60 percent of its capacity. Slowly things were improving as more and more Americans and Northern Europeans felt adequately secure and willing to explore the wonders of the ancient civilizations along the Nile.

And then came 9/11. Though considered a moderate Arab nation, the television pictures of Egyptians publicly rejoicing at the news of the carnage at the World Trade Center struck a dagger into the heart of the industry. Not only were people afraid to fly anywhere in the world, but coming to the Middle East voluntarily was unthinkable. A few seasoned travelers would brave the tour. Those who have ventured into areas of instability know that immediately following a critical event may be one of the safest times to visit. There is heightened security everywhere and terrorists often melt back into the human abyss from which they came.

It was such hearty souls who made up the passenger list of the Oberoi Imperial. *To make ends meet, tour groups were combined. Pyramid Travels had merged the English-speaking customers from the United States, the United Kingdom, Germany, and France along with an elderly Swedish couple. There had been a day's excursion with a round-trip flight to Abu Simbel, which lies 200 kilometers to the south of the life-altering Aswān Dam. Breathtaking in scope and majesty, this engineering and archaeological marvel was saved from inundation in the rising waters of the lake created by the mighty dam. Staring into eternity, four world-renowned colossi, each more than sixty feet in height and weighing over 1,200 tons, face east toward Lake Nasser. Originally located closer to the Nile, the Great Sun Temple of Ramses II was painstakingly carved into 44,000 pieces and moved to the current spot.*

Reconstruction took a decade as each piece was seamlessly situated in the relative position from which it came. So precisely was this task carried out, observers could not discern the vanishingly thin lines between stones. Exiting the temple to the north, tourists would find a small door. Upon entering the coolness afforded there, they would be astonished to find the entire mountain hollow. Only the external statues and the burial chambers had been moved. It had to be seen to be believed!

As their EgyptAir flight landed at the Aswan airport, the window passengers might discern the protection afforded the dam. Over three and a half kilometers in length and comprised of 43 million cubic meters of construction material, this symbol of Egyptian-Soviet cooperation was imposing. Though it would be hard to destroy with conventional bombs, snuggled nearly inconspicuously against sand dunes were pods of air-defense missiles. They reminded everyone of the strategic importance of the dam. For thousands of years the flooding of the Nile brought both life and death. Ninety percent of the country's population resided on the 5 percent of the land irrigated by this river. With completion of the Aswān Dam, that flooding would be controlled and the crops would flourish without worry the waters would rise too high and obliterate the ever-growing residences.

Passing Kitchener Island, the guides looked forward to lunch before the afternoon's stop at Kom Ombu, Temple of the Crocodile God Sebekh. Sumptuous buffets are endemic on these floating hotels. With limited activity and excessive food intake, tourists often acquire unwanted weight on even short cruises. Given the disparate nationalities that made up the small group, the chef dutifully saw to it that every palate was accommodated. And multiple desserts were frequently consumed at each meal.

As they approached the rocky docking area, the captain mentally noted that very few vendors were present. Normally tourists were besieged by artisans and tradespeople from the time they set foot on land until the ship was again under way. Usually good-natured, the vendors would offer a variety of statues, paintings, maps, jewelry, and soft drinks.

The locals would ask to pose with tourists. For that and any other task or service performed one would hear the ubiquitous word baksheesh. *The downfall of modern Egypt's economy was the incessant demand to be paid off. Less frequently children would beg for money outright, an effort discouraged by the tour guides. Well educated, the guides believed that supporting begging only increases the cycle of poverty.*[2]

Approaching cautiously, the Voith-Schneider propeller system designed to rotate allowed the ship to revolve into exactly the right position. Over the gray Phonko PFK-5 intercom system the captain gave the word for the lines to be thrown ashore and the gangplank lowered. Once this was done, he took a pot of tea from the hot plate on the bridge and sat down on the blue-fabric-covered bench.

It was a short uphill walk to the temple. The guards seemed not to notice as the tourists slowly disembarked and meandered along the stone pathway. Suddenly, as the last tourist left the ship, three soldiers sprang into action. Swinging their weapons to the ready, they ran aboard the Oberoi Imperial *and captured the startled captain before he could fathom what was occurring. As at Hatshepsut Temple years before, these terrorists had acquired military uniforms and had been posing as the ship came in. Most of the relatively unskilled Egyptian guards had similarly been taken by surprise. A few had managed to slip away. Their* throats slit and blood attracting the ever-present flies, the others now lay in a pile behind some small buildings.

Before arriving at the admission point, the foreign tourists were surrounded. Clumped in small groups, overtaking them was easy. When one brave tour guide protested he was summarily executed with a shot to his face. The tone was set and there would be no further resistance. Within ten minutes everyone was back aboard the Oberoi, the gangplank withdrawn and mooring released. As the passengers were herded onto the recreational deck the ship again made its way cutting slowly through the bits of floating leaves that dotted the light green waters of the great river.

The terrorists, ten in all, were well versed in their operations. They spoke little as each went about his previously assigned tasks. Ibrahim, the apparent leader, took a place alongside the captain. Obviously Ibrahim had trained on cruise ships before, as he seemed to know every aspect of how the ship functioned. For the moment no one would know the ship had been taken over. That would not last long, as there was a time schedule to keep and demands to be met.

Picking up the microphone of the Model 554 single sideband radio, Ibrahim selected channel 2 from the preset frequencies and began hailing anyone who could hear him. "Allah be praised," he began. Comments invoking the will or graces of Allah permeate the culture and are uttered continuously. Unlike most Western cultures, in Egypt, and other Mideastern countries, religion is a core value that is inculcated into the fiber of every child.

Ibrahim asked to be put in contact with the authorities stationed downriver in Luxor and informed them of the situation. He knew that Cairo would be listening, but he wanted to deal with regional forces.

Word of the reemergence of al-Gamaa al Islamiya spread quickly. While this was a problem for Cairo to handle, the U.S. and European hostages would bring help, both wanted and unwanted. At Ministry of Defense headquarters in Cairo, they knew that the United States and NATO would want to intervene. In fact, U.S. Central Command had a small headquarters stationed in the area. That included a special operations element that could coordinate resources and plan a rescue mission. The high-tech weapons they could bring would be welcomed. But this was in the heart of Egypt and the outcome meant more to them than any other country.

Ibrahim soon sent in his demands. They included release of prisoners in jails around the world, including Egypt, Spain, France, Germany, and Saudi Arabia. Of course detainees in U.S. custody were prominent on the list, the composition of which made it clear this action had been coordinated with outside help, *probably Hamas. As prisoner releases were announced, hostages from that country would be freed. The Saudi and Spanish prisoners were tied to the American response. Any attempts to board the* Oberoi *would be repelled. Further, they had rigged charges and would blow up the ship if a rescue attempt was detected. Ibrahim made it clear the deadlines would be short, but they had sufficient food and fuel to last a long time.*

Cutting power and trading food for hostages is standard practice in such situations. The difference here was that the cruise ship was a self-contained island. They were not dependent on outside resources for anything. The ball was in the Egyptians' court.

Within the hour at Fort Meade, Maryland, NSOC, the National Security Agency's operations center, received a report indicating an intercept had been made between the Oberoi Imperial *and Beirut. The transmission had been short, but it confirmed international cooperation between terrorist groups. Quickly the information was passed to the National Security Council, CIA, FBI, and State Department. At the Pentagon the information went to the National Military Command Center and was passed to both Special Operations Command at Mac Dill, and the Joint Special Operations Command (JSOC). One lesson that had come out of 9/11 was the need for better cooperation.*

In Cairo the American ambassador called for an urgent meeting with the Egyptian president. One would offer condolences, the other military assistance. It was an offer that could not be refused. After all, American foreign aid constituted a large portion of the money necessary to run this country. While Egypt is a democracy today, it is relatively inexperienced in conducting internal affairs. Traditionally a conquered land, in the past two millennia Egypt has only been free for one hundred years. It takes a long time for governments and social structures to mature.

Both British SAS (Special Air Service) and American SOF units were available. As citizens from both countries were hostages, these nations offered to play an active role in any rescue mission. However, Egypt had more than national pride on the line. With their tourism industry already staggering, it was important that they demonstrate the capability of providing adequate security. As a minimum the operation had to be seen as jointly conducted.

Further, they took a hard line and indicated that none of the terrorists would leave the ship alive. Eternally politically correct, both the United States and United Kingdom insisted that efforts should be made to capture them alive. They would be of some intelligence value and their public trial would help gain more support for the continuing War on Terror. If there was armed resistance, killing was necessary—but no executions. The normally placating Swedish government had already signaled that they would act as negotiators in return for release of their hostages.

The Oberoi Imperial *could be visually tracked from the highway that runs along the east bank of the river. Laden with optical sensors, the HMMWV had to make its way from Cairo to Luxor as quickly as possible. However, the self-organizing traffic in the capital is nearly impossible to navigate. The few traffic signals that exist are blatantly ignored and there is no enforcement of any regulations. Worse, many of those aiming vehicles are illiterate and never learned to drive except by trial and error. Forget about safety standards for the vehicles. Travel guidebooks recommend that if foreigners do drive in Egypt they also pray—a lot.³ Arrival at one's destination is again considered the will of Allah, just as is failure to arrive.*

To assist on the first leg of the journey a military and police escort would help clear the path. With their help it would be faster to drive than to lash down the truck on a C-130 and fly it into position. Then they would drive south from Luxor until they spotted the vessel. Once eyes on target were established, they could relay critical information to mission planners. They needed to know how many terrorists there were, where they were deployed on the boat, and, very important, the dispositions of the hostages.

A call was placed to Pyramid Tours. They were asked to provide pictures of all of the employees of the Oberoi Imperial. *As a security measure Egypt required all employees of hotels, cruise ships, and key tourist locations to have picture identity cards. Copies of those photos were kept on file. The system was far from perfect, and cruise captains routinely hired undocumented workers. It was cheaper.*

The intent of asking for pictures was quite simple. The Americans had face recognition programs for their sensors. Subtract the known hostages and employees and the remainder should be terrorists. The plan was not perfect, but it could help prevent innocent people being shot.

Eager to end this as quickly as possible, the Egyptian hostage rescue team was getting ready to deploy. The bridge at Edfu connecting the highway to the Temple of Horus, with its great stone bird, would be passed before they could get into place. However, there was time to get spotters on the bridge to watch how the terrorists went about the passage. It would be on the eastern side, as the water is deeper there. Of interest was where the terrorists placed themselves during the crossing.

The next logical location would be the bridge at Esna. The clearance from the top of the barge to the bridge was such that commandos could fast-rope and be on board in a few seconds. Speed would be essential, but that might not be enough if somebody had a hand on a triggering device.

At Edfu the terrorists had the captain pass between the second and third pillar, relatively close to shore. Four terrorists with AK-47s took up positions along the starboard side of the main deck. Above them on the recreational deck two more were located where they could see the bridge above and the shore to the right.

To the east the sensor operators on the HMMWV could easily observe the decks. While very careful during daylight hours, the terrorists relaxed some at night. They did not realize that powerful infrared thermal sensors were so sophisticated that they could spot gaps between the terrorists' teeth when they smiled at night. The HMMWV still had an escort. Their job was to leap ahead and pick locations providing both clear visibility and concealment.

To everyone's surprise, the next day the Oberoi Imperial turned and headed back upstream. That would throw the rescue planners into high gear looking for alternative interception points. The sensor operators had another trick up their sleeve. From a case they pulled what appeared to be a rather large white gull. Not a bird at all, it was a UAV designed to look like one. It even flapped its wings and had aerial characteristics similar to those of a bird. Close-up it was clearly inanimate. But in flight it could pass.

In the belly of the bird were five small acoustic sensors, each with limited mobility and the capability to find an obscure place and hide. It was hoped that by flying the UAV near the ship they could drop the sensors in places where they might pick up conversations. The signals were very low-powered but easily detectable from the HMMWV stationed not far from the shoreline. From there the messages would be radioed back to Cairo for translation. In the sixth package there was a transponder. This was to ensure that the Oberoi Imperial would never get lost during the operation.

Later that afternoon the ship turned once again and continued heading north toward Luxor. Ibrahim was growing impatient but had been told an exchange might be worked out. Indeed, the Swedes were acting as negotiators but unaware of the rescue mission plans. They could speak in good faith and buy critical time.

Observers had noted that on occasion small wooden boats would venture out from tiny villages along the shore. With their flat wooden oars they would make their way toward the cruise ship. Before closing they would be sent off by stern calls from those on board.

Periodically an army patrol could be seen watching the ship. The Egyptians knew that some response must be visible so as not to raise the terrorists' suspicions even further. A small boat was prepared with special UUVs. These were placed *under piles of reeds. Late that evening, the boat was rowed to a position upstream from the* Oberoi Imperial.

As darkness impaired observation, two SEALs, dressed as local fishermen, dropped the UUVs over the side. Each had an acoustic sensor tuned to the frequency of the Voith-Schneider propeller systems. It was hoped the UUVs dispersed would be attracted to several of the big propellers. Once in proximity of the rotating units they would release their cargo of high-strength propeller-fouling chords. Once firmly entangled, the Oberoi Imperial *would become hard to steer and easier to board.*

At about 3:20 A.M. the captain noticed the small fog bank they were approaching. It seemed a bit unusual as the temperatures didn't seem quite right. Ibrahim, too, saw the fog but missed its significance. To him it would provide cover from the peering eyes he was certain were following them. Earlier the previous afternoon he had caught sight of glint reflecting off the windows of the HMMWV as the sun sank to a low angle in the west. While he guessed someone would be watching, this was the first time he positively identified them.

They were now approaching the bridge at Esna, and the fog might make passage safer, as nobody could shoot at them in it. The passengers were all in the dining area. Most of them were asleep, nearly exhausted from the mental tension of being held hostage. As they were confined in a common area, it only took two guards to watch the entire group.

As the Oberoi Imperial *entered the fog bank and Ibrahim made his scheduled transmission he knew something was wrong. Flying above the cruise ship was an airborne jammer effectively enforcing a cone of silence. Trying the standard frequencies of the single sideband radio, he realized he could not talk to anyone. Quickly he went to alert his troops to resist a pending attack. He had taken no more than three steps when his knees buckled and he collapsed on the deck. He felt a distinct jerk as the UUV wrapped themselves around three of the four propeller systems and immobilized them.*

Looking back, Ibrahim saw the captain, too, was now slumped on the bench behind the steering wheel. As the fog permeated the ship's ventilation system, one after another of the crew and terrorists slipped quickly into unconsciousness. The agent was an improved version of the M99 gas that had been used to free the hostages in Moscow in October 2002.[4] Drifting helplessly, the next thud was the Oberoi Imperial *crashing into a bridge abutment. As it did so the cruise ship spun crossways and commandos dropped silently onto the deck. Just before the attack they had taken the antidote for the incapacitating agent that had been embedded in the fog. In addition, they wore respirators so that they could breathe freely.*

The Americans and British SAS landed first. Weapons ready to exterminate anyone resistant to the drug and prepared to fight, they quickly secured the ship. Medically trained personnel began administering the antidote to the easily identifiable hostages. Learning from the operational errors in Russia, both plenty of medical personnel and the necessary opiate antidote were on hand. Two-man teams swept the ship locating the Egyptians on board. Comparing faces to the sensory database, they began the revival process for employees. Terrorists were bound and then inoculated.

As the search process evolved, the Egyptian hostage rescue team dropped onto the upper decks. Each had a clearly identifiable infrared mark on his uniform. They did not want to have any accidental friendly casualties. It would take a few minutes before the effects of the incapacitating agent wore off. By then the American and British soldiers had dropped over the side into the Nile. It was only a short swim to the waiting boats that would whisk them downriver. In covered lorries they sped through Luxor to aircraft with engines running and the HMMWV already lashed down. Before dawn they were out of the country.

The next day the hostages gave accounts of awakening to find the elite Egyptian unit had saved them. Most of the hostages felt a little ill but did not even know they had been drugged—twice. The press wrote glorious accounts and heaped great praise on the rescue unit. A few investigative reporters wondered how a Third World country could pull off such a spectacular operation. Of course there were rumors, but as usual, nobody at Fort Bragg was talking.

The terrorists did not fare as well as the hostages. Six months later in Cairo they were placed on public trial. All were convicted. As everyone realized the damage the terrorists had done to the country's economy and wanted to send a strong message of deterrence to others, all were sentenced to death. Death by hanging is carried out in Egypt. Some still say the terrorists accepted martyrdom with dignity and that would encourage other dissidents.

ENCAPSULATED IN this chapter are several critical issues. The ambush at Luxor's Hatshepsut Temple on November 18, 1997, was not about killing tourists. It was a blow designed to cripple Egypt's struggling economy. Years after the attack, tourism has not returned to the preattack levels. This chapter suggests that the actions of a few terrorists would have devastating consequences, ones that could destabilize the current government and shift power in the region.

Hostage rescue operations on ships and boats are very difficult. The weapons used in this recovery do exist, albeit not all of them in the U.S. inventory. The use of chemical incapacitating agents is very controversial. The chemical warfare purists believe their use would violate existing treaties. As a pragmatist, I believe we signed these treaties for altruistic purposes, never considering that advanced technologies would provide the capability to use chemical agents in a life-conserving manner. Just as we would not set speed limits based on automotive technology and highway engineering of the early twentieth century, future treaties based on chemistry of that era should not bind our options.

CHAPTER FIVE

PORTS OF CALL

SHORTLY BEFORE 11:15 IN the morning, Seaman Raymond Mooney spotted a small boat closing on the destroyer. He thought little of it, as the busy harbor contained many such craft zigzagging about on their daily chores. As the destroyer had just begun a refueling operation, his job was to watch for spills, not guard the ship. Besides, the two men standing on the boat were smiling and waving as they came immediately adjacent.[1]

AT 11:18 A.M. on December 12, 2000, an explosion rocked the USS *Cole* as it lay at anchor in the steamy port of Aden. An ancient Arabian port, it has been known to harbor pirates, thieves, and terrorists for hundreds of years. The attack inflicted heavy casualties, seventeen dead and thirty-nine wounded. The enormous explosion did substantial physical damage and left a gaping forty-by-forty-foot hole in the hull, causing the ship to list dangerously. However, this blast not only shook the 355 members of the 505-foot Arleigh Burke–class guided missile destroyer; it also reverberated throughout the entire U.S. Navy. Force protection was catapulted to the front burner.[2]

One of the missions of the U.S. Navy is projection of power, which is sometimes accomplished by showing the flag in foreign ports. Usually this demonstrates friendly international relations with the country visited and offers a chance to replenish supplies. However, these port calls present increased vulnerabilities and a complex set of diplomatic issues. At sea the Navy is responsible for its own security and implements complex defense in depth. The fleet can control the surface waters for appreciable distances and have air cover above and, if necessary, submarines below. They can dominate their environment.

When ships are in foreign ports, however, security is the responsibility of the host nation and one side of the vessel is contiguous to the pier. Most likely other foreign vessels are in close proximity and the U.S. naval ship is vulnerable to land-based attacks that can be initiated from very close range. As a result of the attack on the USS *Cole* a Department of Defense commission was appointed and provided recommendations to enhance antiterrorist/force protection capabilities.[3] They recommended a three-tier defense that still has problems with assaults initiated from the land where ships are docked.

In many ports, a large number of commercially available personal watercraft (PWC) occupy the waters near the dockage where the Navy's ships are at anchor in a harbor. Intrusions into the space that is "unacceptably close" to a ship may take place by innocent civilians, such as fishermen or pleasure boats. In many areas of the world there are protesters in opposition to American policies in general or a military operation in particular. Such ostensible protest demonstrations may offer effective cover for terrorists. As in the case of the *Cole,* hostile intent was not evident until the small craft blew up.

The procedures emerging within the AT/FP initiative involve both lethal and non-lethal weapons and procedures. When entering the port of a foreign country, a U.S. Navy ship goes through a substantial—and growing—series of checks to assess the risk of terrorism and threats from hostile individuals or groups. In fact, these begin long before the ship arrives and include coordination of both U.S. and foreign intelligence services. Then a force is dispatched to jointly sweep buildings close to the port, assure the necessary security operations have been taken, and enhance security through acceptable means.

The primary force protection architecture for ships in ports involves layers or zones related to the nature of a potential threat. Any approaching watercraft could pose a potential threat if it were to come within a specified distance of the ship, that is, within "Zone 1" (or the "outer defense bubble" in three dimensions). At such a point, action is taken to warn the approaching craft to keep its distance and to determine if such an approaching vessel is either hostile or unaware of the fact that it is a potential threat to the ship. Warning signs, lights, audible warnings (sirens, horns), and other perimeter indicators are appropriate at this distance. If the vessel approaches closer, within Zone 2, a series of non-lethal warnings or actions may be employed. These non-lethal actions should be sufficiently intense so that any innocent intruder entering Zone 2 clearly recognizes his vulnerability to attack if he continues closer. If the approaching craft enters Zone 3, then the intent of the approaching craft can be assumed to be hostile and lethal actions may be employed.

A high level of uncertainty exists in situations where a vessel is in Zone 2. The rules of engagement (ROEs) specify that every commander has a right to defend his/her own ship, including lethal means when necessary. Due to the short period of time during which a high-speed PWC may approach a stationary vessel, the ROE must include authorization for the individual on duty to take the necessary action. If, for example, an approaching vessel is traveling at 90 knots from 2,000 yards to 1,000 yards (notionally, from Zone 1 into Zone 2), a sailor charged with ship security has ten to fifteen seconds in which to decide and act. He must have a clearly defined set of actions related to increasing levels of lethality: (1) assess, (2) warn, (3) threaten, (4) intimidate, (5) incapacitate (personnel or matériel), (6) disable, (7) damage (matériel), and finally (8) destroy.

Non-lethal methods allow the sailor to address steps 2 through 7 before lethal procedures are required. Crews must be properly trained and have the authority to take action in sequence to a complicated and rapidly changing scenario.

Actions in outer Zones 1 and 2 that may be taken as a result of the AT/FP initiative include the following:

Detection. Detection of vessels that intrude within Zone 1 is obviously necessary. Sentries, EO devices, radar, and sonar may be used extensively to determine the presence of objects within a designated distance. New detection devices may be appropriate to enhance the ability to detect small boats, subsurface swimmers, approaching small aircraft, and intruders in land vehicles or on foot. However, it is not a simple matter of detecting a specific sound or device. In ports there will be many legitimate objects in the general area. New systems must identify and locate the source of that object. Based on the activities of the object a determination can be made as to whether hostile intent is indicated.

Attention-getting actions. Signs indicating the limits of a perimeter defense may be placed at appropriate locations to warn approaching vessels. Horns, sirens, or lights may be exercised to gain the attention of the approaching vessel. Sentries may attempt to provide warnings, and animated warning signs may be triggered. For subsurface swimmers, acoustic warnings can indicate that they should not approach any closer.

A useful tool is the handheld LE Systems' Laser Dazzler that has the ability to create a ten-meter-wide cone of light at a distance of

400 meters. The pulsating green light is too brilliant for a person to look directly at it. This small light also serves as additional illumination when used at night.*

Non-lethal actions with lethal weapons. It has been established that a show of force can cause an adversary to back down. A clearly overt action is to fire a shot across the bow of an approaching vessel. This is a "non-lethal" action that has been recognized for centuries and represents a recognized signal that if no change in course is taken, a lethal shot may be fired. Besides endangering other activities in the area, such action would be deemed extraordinary and could be offensive to the host country providing port security. Alternative actions involving non-lethal procedures are highly preferred and are being developed for this purpose.

Riot control agents. A variety of riot control agents (RCAs), such as rubber pellets/batons, water cannons, flash bangs, pepper spray/balls, and other chemical lacrimators/irritants, may be used under the restrictions of the Chemical Weapons Convention. Combinations of agents can also be considered (e.g., mixing lacrimators with water cannons). The delivery of RCAs to small vessels with remotely piloted watercraft or UAVs is appropriate for the distances and situations anticipated by such encounters and can be executed at ranges of over 1,000 meters.

Detection of offending materials. If a vessel is to be boarded, either in the AT/FP scenario or in enforcing sanctions, detection to determine the presence of offending substances on board may be necessary. Chemical sniffers capable of detecting the presence of explosives or other substances are available. Improvements in placing them in very small devices are under development. Current methods of detection are very short-range. This requires some sort of remotely controlled platform to get the sensor close enough for detection. A small UAV may be used to fly close enough to collect air samples for real-time analysis. Another approach will be expendable sensors that can be dropped onto the craft and report chemical analysis via radiotelemetry.

Vessel stoppers. Vessel stoppers, such as the running gear entanglement system that has been previously described, will soon be available. Exhaust stack blockers have been evaluated. They were deemed not attractive due to the difficulty of placing the blockage in the exhaust

---

* A complete description of the Laser Dazzler can be found in chapter 1.

stack of the vessel. However, they can be designed to fit in UAVs that actually fly down the exhaust stack instead of attempting to hover above it. The biggest concern after delivery is causing excessive pressure in the engine and accidentally blowing up the ship. Casting a net across the bow of a vessel (sea anchor vessel-stopping system) has been suggested; the net is attached to parachute-shaped "drogues" that open and impart considerable resistance to the continued motion of a vessel.

Delivery of this system or RGES remains a challenge. Remotely piloted small craft have been tested to assess their ability to perform this function. An alternative proposal is a small-craft disabler, which inserts a spear into a hull at the waterline and deploys a fin that drags in the water, making steering impossible.

Surface patrol vessels. Small-craft patrols operating in conjunction with the docked ship allow closer monitoring of any vessel entering Zone 1; they are being proposed as standard operating procedures for vessels in port. The delivery of warnings, vehicle stoppers, or other items for which a close approach to the offending vessel is required may represent a substantial challenge. The Coast Guard reports that high-speed personal watercraft and other high-powered vessels frequently outrun pursuing Coast Guard vessels. A relatively inexpensive Jet Ski to which a remote control and monitoring system may be attached is commercially available at an estimated cost of $50,000. Speeds of up to ninety knots in calm water are advertised. This Roboski has successfully deployed an RGES to stop a boat. Costs of the Roboski/RGES system are $87,000. A Roboski platform might also be used to deploy a drag chute over a vessel; warning devices such as sirens, flashing blue lights, and strobe lights; flash-bang munitions; pepper spray; blunt trauma munitions; or a water cannon. It might also be used as a ramming device. These ideas are under consideration for countering the threat of a high-speed intruding or escaping vessel.

Unmanned aerial vehicles. A variety of UAVs have been suggested for patrolling airspace. The one recommended to the navy weighs thirty pounds, has a ten-foot wingspan and thirty hours' endurance, and costs $8,500 with a data link. This would allow for continuous observation of the area of interest. If small vessels are approaching, the UAV can be directed to monitor them closely and send back pictures of what can be observed above the decks.

Smaller UAVs have also been proposed for close-in protection. These can be equipped with chemical detection sensor systems. Flying very close to the vessel in question, the UAV would scoop up air samples and determine whether the boat was carrying a large amount of explosives.

The low-cost expendable electronic warfare (EW) killer (LEWK) is another vehicle that can be launched from various guns or launch systems to provide over-the-horizon monitoring. This UAV is much larger than the others and can fly out at greater distances while the ship is at sea.

The preceding measures address hazards that may approach on water. Additional threats may appear with divers, swimmer delivery vehicles (SDVs), or unmanned underwater vehicles. Sonar monitoring can detect the presence of such threats. Several experimental systems with the designation ANW QX2 have been introduced to selected ports to monitor such activity using active sonar. Actions to minimize risk if such intrusions are detected include: (1) a loud sonar blast sufficient to cause a swimmer to surface or to break eardrums; (2) counterdiver activity with people or marine mammals; and (3) in extreme cases, subsurface munitions (lethal means) i.e., mines.

CHAPTER SIX

ANOTHER WORLD

PASSING THE PORT CITY of Iquitos, the long, low aluminum boat continued northwest from the Amazon River onto Rio Nanay. As they crossed the ripping waters of the merging rivers three fast-moving boats from a drug interdiction unit met them. The unit was stationed just to the east of the city of 400,000 inhabitants on the north bank of the mighty river. Well-funded at over $77 million from the United States and run under DEA guidance, these operations frequently patrolled the rivers hoping to find cocaine smugglers.[1] However, the local passenger vessels were normally greeted with waves from the heavily armed counternarcotics force.

The Amazon area of Peru is exceedingly poor and indicative of the anachronisms typical of clashing cultures. The tribes people, such as the Boras and the Yagua, are barely one step out of the Stone Age. Those who have not migrated to Iquitos are transitioning from hunter-gatherer societies to humble agrarian beginnings in which they aggregate into more stable communities. However, many young people from the tribes have entered the city in search of the meager jobs available. Poverty in the once-thriving community is endemic and on a par with that of Bangladesh. More than one hundred years ago the opulence brought by the once-thriving rubber industry suddenly dissipated as British smugglers transplanted the precious trees to Malaysia for ease of export.[2] Even now, heavy goods are accessible only by sailing more than a thousand miles up the treacherous Amazon River on oceangoing vessels.

Fewer than 5 percent of Peru's population live in the Amazon Basin, and there are no roads in the north that cross the forbidding Andes Mountains, with peaks rising to nearly 20,000 feet. There probably never will be, as the Quechua and Aymara cultures that dominate the country live in the high mountains and along the Pacific coast. If asked, most of these primitive tribes people would identify themselves as Amazonians, not Peruvians. In return, they are at the very bottom of the food chain when it comes to social programs. Response to 911

105

calls in these remote areas is measured in days, not minutes. These people, as UN Secretary General Kofi Annan remarked at the turn of the millennium, are part of that half of the world that has never made or received a phone call.

Transportation between towns and villages is almost exclusively by boat. It is not unusual to see small groups of people standing on riverbanks patiently waiting to flag down a passing craft. Experienced pilots navigate through the ever-changing web of islands and channels that comprises more than 8,600 miles of tributaries to the Amazon in Peru. With each rainy season entire islands are washed away and new ones formed at other locations. Boaters must be on constant lookout for both debris and the large logs that are floated downstream to lumber mills.

Contrary to popular belief, the natives have no word for jungle. There is only the forest. That is differentiated into the low forest, which is submerged during the rainy season, and the high forest, in which they may live year around. Trapped in a living paradox, the natives do the best they can. Health care has improved, thus leading to increased population. That further stresses the resources available from the fragile environment that has supported their way of life for centuries. Hunters continually must go deeper into the forest in search of withering game supply. Others cut more of the forest hardwood to be made into charcoal for Iquitos. Yet despite these unbearable pressures, the people maintain a propensity for effervescent happiness.

There are two distinct groups of foreigners in Iquitos—tourists and narcs. The easily discernible tourists, on the one hand, herd around ogling the panoply of native crafts available at inflated prices. Based on their naïveté the tourists, awaiting their brief voyages on air-conditioned cruise ships, are called fresh meat by the vendors.[3] The narcs, on the other hand, are conspicuous by their military-like demeanor and can be found congregating in the bars along Malecon Tarapaca. There they freely partake of the affections lavished upon them by young females seeking to improve their lot in life. Needless to say, the narcs have brought a dramatic rise in venereal disease not unlike their counterparts stationed in Third World countries around the globe.

Located close to both Colombia and Brazil and far from Lima, this area is frequently used for transporting cocaine products. For several years the United States provided airborne over-watch for illegal flights, and between 1994 and 1997 about twenty-five drug-smuggling aircraft were shot down by the Peruvian Air Force.[4] Then, on April 20, 2001, the unthinkable happened. A U.S. reconnaissance plane vectored a Peruvian plane to intercept a Cessna 185 that they could not identify. It was about ten in the morning when the Peruvian Air Force plane strafed the intruder, instantly killing missionary Veronica Bowers.

A bullet passed through her body and struck Charity, her seven-month-old daughter, in the head as she lay resting in her mother's lap. Despite being shot through the legs, the pilot, Kevin Donaldson, miraculously was able to crash-land the crippled plane on the Amazon River near the small town of Pebas.[5] Though the suspicious aircraft was clearly disabled, the Peruvian plane continued to machine-gun the survivors as they swam for safety.

If it weren't for periodic tragedies, Peru would never make international news. This incident reverberated loudly and brought about revised procedures. Contrary to initial reports, it was learned that Donaldson had filed a flight plan and was in radio contact with the Iquitos airport when they were shot down. Condolences abounded, but after public memory faded, in January 2002, the ill-conceived airborne interdiction program resumed.

Since November the agents had been attempting to find a coca paste processing area located somewhere in the forest north of Iquitos. In the dry season of December the receded waters exposed large areas of land, and a temporary facility had been established. Being situated on major water sources and only two degrees off the equator means constant heat and high humidity. Therefore, broken clouds are nearly always present except for periods of intense rainfall. Thus satellite photo coverage is often inhibited. But the three steps in cocaine processing usually took place near a river, possibly with an airfield collocated.[6]

In Iquitos natives were showing up with modest amounts of money. A pittance by Western standards, it was still enough to raise suspicion in local authorities. They soon learned that certain tribes were being offered jobs dancing in the coca leaves. Dancing is a nasty process that entails tromping barefoot on coca leaves in vats of kerosene to produce the coveted paste. Chewing the leaves as a stimulant, the natives can dance for hours. Since even the concept of OSHA is inconceivable, no concern is paid to the health of these indigent workers. If they complain there are many more to take their place.

Issues of morality are beyond comprehension to these humble peasants. Dancing may be the only cash-producing job that is offered to them for years at a time. Living in the forest or in remote villages, they could not fathom the consequences of the product they help make and barely benefit from. They, too, are but a resource to be used and discarded.

Once the general coca-processing area was identified, the patrols began looking for covered wooden boats that might be carrying kerosene drums or diluted sulfuric acid there. They would stop the heavily laden banana boats as they skimmed along the rivers, gunnels barely an inch from the waterline. Finally a break came when drums of the precursor chemicals were discovered buried under thatching material. The fisherman was trapped between a rock and a hard place.

The Peruvian agents would probably beat him to death if they did not get adequate information. If he was lucky and survived, while serving time in a prison at 12,000 feet in the Andes was tough duty for anyone, as he came from the Amazon, the constant cold temperatures and reduced oxygen in prison would be nearly unbearable. Conversely, if the narcoterrorists learned of his betrayal, a very unpleasant death was assured for him and his family.

Yielding to the most impending danger, the fisherman decided to talk. There were no words to accurately describe where the camp was located. He could, however, lead the agents to the laboratory and would do so for his freedom. While the fisherman was in no position to bargain, the DEA had little interest in such small fry and knew the dilemma he faced anyway. They agreed to a plan in which agents would follow him until they were in the vicinity of the camp. He would continue alone, drop off the supplies, and be on his way. If the agents learned, then or later, that he had tipped off the coca processors, they would simply kill him. The biggest issue was developing a cover story to explain his delay in returning to the laboratory.

Disguised as native fishermen, the Peruvian agents stopped following within a mile of the site and allowed their boat to drift. With darkness nearing, they eased their way to the shore and made a quick reconnaissance. Certainly there would be observation posts along the river. At dawn the following day the agents spotted a small fire on the western bank. From their hidden vantage point in the woods they watched as two lookouts went about getting ready for the day. It seemed clear that they did not think the camp had been compromised. Rather than risking discovery, the agents retreated and made their way back to where they had stashed their fishing boat.

Using GPS the agents had determined the location of the point of entry. Due to the cloud coverage and changing topography, accurate maps of this area just don't exist, but the satellites never lie. Next would come the task of pinpointing the laboratory and devising a plan to attack it. The raid had to come as a surprise, but getting close without detection could be a problem.

Two days later a recon team returned to their hidden site by boat. Any aircraft activity in the area would be suspicious, as private planes rarely fly over the area. Instead the recon team brought along petite transparent aerostats. Two hundred feet above the ground they became nearly invisible. However, care was necessary, as the forest natives have a far greater sense of environmental awareness than most people do. Therefore, the infrared sensor package was lofted in the evening and retracted before dawn. Changing locations and triangulating, the team finally located the facility about a kilometer farther upriver.

For two days the team stealthily observed the area. They learned that the camp was on the ground about thirty feet above the river but in an area that was only about one hundred meters wide. They noted what appeared to be a nest of logs and flotsam that was not moving downstream. At night one team member swam out to find the jam was constructed to prevent any rapid approach by their fast boats.

At night they were able to come very close to the barracks area. There they overheard conversation among workers, narcoterrorists, and a third group. Providing security for this operation were guerrillas, remnants of Sindero Luminoso. Though former President Alberto Fujimori had severely damaged their infrastructure, hard-core groups of these Maoists still remained active. Providing security for the drug trade was a natural fit to acquire much-needed cash. Pressure on Sindero had diminished slightly after Fujimori fled to Japan and abuses by the intelligence services came to light.[7]

While laboratories were normally defended, the presence of the Sindero Luminoso guards meant that they probably had more automatic weapons than are routinely encountered in drug raids. Plans would have to be made for suppressive fires to protect the assault force during the initial phase of the operation. Given the minimal cleared area, the fight would likely be at close quarters and preventing agent casualties from friendly fire a major concern. As the old saying goes, friendly fire isn't.

As a precaution the agents obtained samples of the water. Never environmentally conscious, the people who operate the drug laboratories dump the residue directly into the nearby rivers. In fact, several tributaries already have become totally contaminated, resulting in fish kills further exacerbating the plight of the forest people who forage there. Analysis of these samples would confirm the type of processing that was occurring at this laboratory.

With the War on Drugs, there were still rules of engagement and strict guidelines for use of U.S. forces. Despite the presence of the guerrillas, ordering up C-130 gunships for fire support was out of the question. In Washington, drug processing in the Amazon was considered an annoyance to be raised only when it was politically beneficial. The potential of losing a valued aircraft, even by accident, did not warrant creating an international incident. Better for a few agents on the ground to buy the farm. That rarely would make a ripple on Capitol Hill or with the media.

It was determined that a single aerial photo mission would be conducted. The small UAV with IR cameras would be launched from a clearing about three miles from the laboratory. As the prevailing winds were westerly, the UAV would circle widely to the north and west at an altitude of about five thousand feet. Once in position it would be flown to the east just above the northern edge of the laboratory. When it was still several hundred meters out from the objective, the baffled engines would be cut, allowing the UAV to drift silently to the target area. Photos taken and safely downwind, the engine would be restarted and the UAV flown to the retrieval point.

Back at the DEA camp the images would be entered into a computer, allowing for a three-dimensional model of the site to be generated. This provided the special forces trainers the information necessary to rapidly construct a near full-scale model of the laboratory and surrounding area. While the trainers could not accompany the operation, they could see to it that this high-tech simulation capability could help prepare the raiders.

The raid would require a combination of human ingenuity and cunning supported by brute force, albeit limited by U.S. standards. To ensure maximum news coverage and political benefit, it was decided that the raid would take place in the middle of the week. A small team would infiltrate the area on Monday. They would locate the lookout posts and prepare the artificial dam for demolition. Their intent was to blow a hole near the center of the blockage so that the fast boats with their machine guns could quickly dash upstream and suppress any fixed weapons sites.

The main raiding force would be flown directly into the laboratory area on U.S. purchased Black Hawk helicopters. Their supporting fires would be supplied by the AH-6J, somewhat equivalent to the U.S. Little Birds of the special operations forces sans some sophistication. However, their 7.62mm miniguns and 2.75-inch rockets would do the job and they could hover in for very close coverage.

Effective immediately upon beginning construction of the target model, the main camp was closed down. No one was allowed to leave and the phone lines were cut. The raiding teams quickly began preparation under the tutelage of both the U.S. Army Special Forces, who would stay behind, and paramilitary personnel, who would accompany the raiding force.

Inside a sequestered building in the DEA camp, radio intercept systems were turned on. The Peruvians were not aware of this capability, but it would warn U.S. operation leaders if someone was attempting to get a message out. If needed, a jamming system was prepared to block all radio signals in the area. In addition, the radio stations from Iquitos were all monitored for any sign that a message warning of the attack might be transmitted. As an added precaution, word was distributed locally that several of the agents had come down with a fever of unknown origin and so the base had been closed as a medical measure. Further information about the nature of the disease would be disseminated as soon as Centers for Disease Control personnel arrived.

That night after midnight the advance party slipped out of the port and began the trip upriver. They moved deliberately, using IR lights and night-vision equipment to avoid the ever-present debris. Before dawn they had docked at their old location and searched carefully for any sign that it might have been discovered. The wooden boats, now carrying explosives and weapons, were snugly secured and camouflaged.

The symphony of sounds emerging from a million birds and insects announced the coming of dawn. Care was taken to allow the orchestration to proceed, for the forest has a way of alerting indigenous inhabitants to impending danger. Shrill cries from monkeys or an unexpected flight of birds could alert the drug manufacturers. Therefore, the team members lay motionless through most of the day while carefully noting everything around them. Save for highly skilled snipers, few Westerners have the patience for such intense attentive inactivity.

Throughout Tuesday the team noted the locations of the downriver lookouts. Late that night they returned to the river and picked up the prepacked demolition kits that would be used to blow the makeshift dam. Two cuts would be needed to provide an opening for the gunboats. Traveling by boat was too risky, and the current was too strong for a swimmer. Therefore, the team had been provided with two commercially available Torpedo 3500s, small electric diver propulsion vehicles. They were both quiet and sufficiently powerful to move the sappers and their cargo to the target.

At exactly 2:00 A.M. Wednesday the six fast boats pulled slowly out of the docks at the DEA base. As they approached Iquitos two left the formation and went to the western bank where the Peruvian Air Force keeps their pontoon-rigged planes. Then turning upriver they went into the pier at the local yacht club. The two-story building has a commanding view of the river and stores the boats owned by local dignitaries. While most people in the area were very poor, some had managed to attain considerable wealth. The yacht club officially would be closed, but sometimes the members chose to party at the exclusive site. Placing a small group there ensured that the other boats had not been spotted and no one would attempt to radio an alert.

The underwater demolition team arrived at the dam and placed charges designed to produce a wide hole allowing the fast boats to enter. Once implanted the trigger mechanisms were activated. In this case they were receivers tuned to a specific frequency with a coded signal required. This would allow the raid commander to detonate the charges on his command. Therefore, he did not have to be concerned about premature explosions should the flight be delayed by unforeseen circumstances.

That task completed, they took the Torpedoes to shore and hid them. Then they drifted silently downriver toward the lookout sites. There were two, one on each side of the tributary, spaced about five hundred meters apart. Small cooking fires dimly illuminated the area, making it easy to locate the sites. The guards used coffee to stay awake late at night and had taken precautions to shield their fires from anyone traveling upriver. They did not expect an attack from behind them.

At 4:15 A.M. the helicopters lifted off from the Iquitos airport and flew the short distance to the Peruvian army base on the main road. There they touched down briefly. To the astonishment of the soldiers waiting to board them, they immediately-departed heading to the south. Part of the cover story was that an army training exercise would be conducted Wednesday night. The officers and troops actually believed that story and had planned to be airlifted to a forest area to the south. Inevitably word of this had spread through the local community.

Traveling briefly to the south, the entire flight turned east and followed the Amazon to the DEA base. Landing, they were quickly refueled and boarded by the raiding party. Five of the six Black Hawks contained ten raiders each, including an American paramilitary on each bird. The sixth was the command and control ship. The Black Hawks departed at 5:00 A.M., quickly followed by four heavily armed AH-6Js. Initially they flew at 500 feet above the forest but dropped down as they approached the target.

Meanwhile the recon team infiltrated the two lookout camps. Only one man was awake at each site. Before they could react, they were struck by tranquilizer darts. Though armed with silenced pistols, the team was able to bind and tape each of the sleeping lookouts before they could alert another. Once the sites were neutralized, the team turned the fires so that they could be seen from the boats moving slowly toward the dam. Their final task was to establish ambush positions in case someone came to check on the guards.

Rather than risking flying into a tight landing zone at night, it had been decided to attack shortly after dawn. The raiders all wore reflective tape on their uniforms and helmets. Once engaged they wanted to be sure where the friendlies were in a situation that could be chaotic.

At about five minutes after 6:00 A.M. the signal was given for the Black Hawks to dash for the laboratory. Coming up the river, the command and control ship touched off the charges on the dam just as the raiders hit short final. Having had the benefit of reviewing the exact landing sites from the aerial photos, the pilots knew precisely where to land. In less than a minute the lab was surrounded with raiders attacking their rehearsed targets.

The barracks area proved to be the most problematic. About a dozen members of Sindero Luminoso were sleeping with their AK-47s provided courtesy of the former Soviet Union, upon whom Peru had relied for military support in years gone by. The raiders blocked exits to the forest so the Maoists could either surrender or stand and fight. Long prison terms, as had befallen many of their former colleagues, were not an enviable option. They chose to fight.

It was then that the AH-6Js made their entrance. The location of each of the friendly raiders was easily noted. The lead gunship placed two rockets into the wooden buildings. Flame and splinters flew everywhere, but the hardened guerrillas continued to fire in all directions. With miniguns blazing, a team of gunships made strafing runs down the length of the building. Then the fourth ship hovered in close and slowly moved sideways. When the run was finished, firing stopped and the building was hardly recognizable.

Following the explosion, the waiting fast boats accelerated through the gap and dashed toward the camp. A number of the narcoterrorists had bolted for the river. When they saw the helicopters land and block their retreat to the dense forest, they assumed the river might be a way out. They had not counted on the approaching boats that also provided covering fire. Two terrorists did manage to stay far enough inland to be missed by the boat crews. Following the trails toward the lookout sites, they ran smack into the waiting ambushes. They were so shocked that no shots were fired as they were taken into custody.

The raid had been perfectly timed. The laboratory had a large inventory that was due to be shipped out that day, and at about 10:00 A.M. a Cessna was spotted approaching the unimproved airstrip. However, they, too, spotted the smoke that was still rising from the area and veered off. Almost before they could alter their course, two Peruvian jets were on them. They would have lived had they followed the instructions to land.

The bust was huge, with near a ton of cocaine paste confiscated. Alerted to the impending raid the night before, several key Peruvian military officials flew from Lima to take credit for the raid. The American embassy also made news releases and the ambassador decided this would be a good photo op. Even at that level, few understand whether the U.S. military should be involved in these drug operations. Worse, most of the cocaine processing had been moved to the Pacific coastal region, making the expenditure for the DEA camp in the Amazon even more questionable.[8]

As fate would have it, on the same day an Amtrak train derailed on the New York to Boston corridor, causing thirteen deaths and injuring scores of others. It also provoked the closing of the nearby interstate highway. The news pundits pontificated about how poorly the heavily subsidized railroad was run. News of the successful raid in Peru was banished to a small column on page 12.

• • •

THE SO-CALLED *War on Drugs* has been controversial from its inception. Advocates claim drug use threatens U.S. national security. Emphasis on interdiction has expended vast sums of money but failed to lower the best barometer of success—street prices. They have remained fairly constant except for a brief period following 9/11 when border security was extremely tight.

This chapter shows that use of high-tech sensors coupled with aggressive action by drug enforcement agencies can produce periodic minivictories. It is hard for most Americans to grasp the vastness of this area of the Amazon or the difficulty of conducting operations in this watery primeval wonderland. At a time when we ask our armed forces to be able to function anyplace in the world, this examines one of the extensive but extreme environments in which we find ourselves engaged.

CHAPTER SEVEN

HIMALAYAN HOLIDAY

THOUGH ONLY 4 P.M., *the sun had already slipped behind the steep mountain ridge. The December air cooled quickly, signaling it was time to be heading back to the lodge. Sanctuary Camp, located on the Modi River, held a breathtaking view of the sacred Nepalese mountain, Machhapuchhre. Due to its unique geological configuration and rising imposingly over 22,700 feet, it is known to Westerners as Fishtail. It dominates the serene landscape. To the north, at still higher elevations, were the vast, magnificent mountains of the Annapurna Range. Formed by the collision of tectonic plates, their majesty continues to evolve as India and Nepal gradually slide under Tibet to create the most precipitous geographic boundary in the world.*

The winding trail back to Sanctuary Camp, comprised of rough stones, was a challenge when traversed in daylight. Trekking it at night brought two dangers to bear. First, with ease one could twist an ankle or, worse yet, fall and break a bone. In this roadless remote area of the Himalayan foothills, here, too, emergency response is very slow. The only way out for most of the seriously injured from this ambulance-deprived site is in a wicker basket carried on the back of a local Sherpa. The second concern took the form of an incipient but growing insurgency. In this poverty-ridden country, the Maoist guerrillas gained wide acceptance by simply offering something different. It didn't matter that they established themselves on the defeated model of the Sindero Luminoso movement of Peru.

Engrossed in watching the optical illusion of the waterfall, Jim and Gloria Dunlap were blissfully unaware of the foreboding shadows that appeared on the Old Tibetan Highway, which ran a few hundred feet above them. They were content to stare intently at the white froth of the cascading stream. By so doing and then shifting their gaze slightly, they saw the green and brown vegetation and rocks on the right side appearing to move in a counterclockwise pattern

115

upward and across the face of the cliff. Though within a kilometer of Birethanti, a substantial village for this part of the region, they were quite isolated.

"Namaste," the Dunlaps said as they saw three young men approach them. This typical greeting had become second nature as they passed local tribespeople. Normally it was returned with endearing smiles of the gregarious children who inevitably sought handouts. This time, the Dunlaps noted the dour looks but still did not perceive what was about to happen.

Closing within three feet of the couple, the leader pulled out a pistol and waved it menacingly at the now-bewildered couple. He spoke no English but pointed for them to follow. Jim Dunlap, a retired computer consultant from Wichita, assumed this was a simple robbery. While rare, they did occur. Most tourists carried more cash than the average Nepalese would make in a year—if he could find a job at all. Then there was usually camera equipment, expensive watches, and the jewelry that American women can't seem to be without. There was no hesitation as the Dunlaps offered their wallets, watches, and cameras. Jim even produced the anticipated hidden money belt that fit snugly around his waist. Not relinquished so freely was the emergency stash in his ankle pouch.

The money quickly disappeared. However, the Dunlaps were surprised when the bandits pointed at them and indicated they must follow. They also were shocked to see the abject fear in the eyes of GB (the Nepalese often use two letters to abbreviate their names), the porter who had been assigned to watch them. Uneducated but with strong backs and ingratiating smiles, each porter carried the heavy packs for two people and willingly accompanied the tourists as they wondered about exploring the scenery. On the side excursions their main job was to see their charges didn't get lost or injured. Private security was not one of their functions.

Slightly overweight, Gloria was allowed to keep one walking stick with which she could keep her balance. They were pushed roughly toward the rock-strewn roadway, turning left into the mountains and away from the relative safety of the village. Following the gun-toting leader, the Dunlaps obediently fell into line. The terrified porter stumbled uncharacteristically near the rear of the column. As GB was a worthless commodity to the Maoists, his fate was sealed as soon as they had moved about a quarter of a mile farther from Birethanti. There was an abbreviated scream, interrupted as a single swift blow from the trailing Maoist's khukuri decapitated the hapless fellow. It was the two distinct dull thuds as head and body fell apart that caused the Dunlaps to look back. Wide-eyed, Gloria urinated where she stood. Jim's knees buckled as he saw the Maoists push GB's torso off the steep side of the road, then almost carelessly grab his head by the hair and flip it about thirty feet into some bushes.

With steel-cold eyes, as if impervious to what he had just done, the executioner turned and moved up the trail. Emotionally stunned and gasping for breath, the Dunlaps were prodded back to reality as they were again herded north up the trail. It would be several hours before the initial fears of impending immediate execution would subside. Much later that night the truth sank in. The Dunlaps realized they had been kidnapped for ransom.

As the afternoon tea was set at Sanctuary Camp, Sherbahadur Pun, the tour group leader, was becoming anxious. With tensions rising in the area, Sherpa Enterprises, the company for which he worked, had insisted on guides with military experience. The signs of Maoist activities were endemic. There was, however, an established agreement between the insurgents and the tour industry. In return for modest payoffs the guerrillas would leave tourists alone—at least until now. Since the Maoists had broken the short-lived truce in 2001 the Nepalese Army had been applying increasing pressure, and the ability to support their movement was impaired.

Sher, as the Westerners called him, had spent twenty-seven years in Her Majesty's service. Achieving the rank of captain, Sher had traveled around the world protecting British interests with the 2d Regiment of the highly respected Gurkha Rifles. Because of antiquated agreements between the British, Indians, and Nepalese, he received only a small fraction of the salary provided British officers of the same rank although he had performed equal service. It meant that like most loyal Gurkha Rifle retirees, who could not survive on their meager pensions, he had to seek other income. Sher, and many others like him, entered the tourism industry and were reasonably successful as long as the visitors kept coming. Having lived in the West, these former military officers knew the customs and expectations of the tourists. With great effectiveness they met even the most inane requests of the often out-of-shape and elderly trekkers who had sufficient money to employ formal tour companies. Their clientele differed significantly from the near-indigent foreign youngsters who wandered the hills in search of their inner spirits.

As darkness fell and GB had not returned with the Dunlaps, Sher knew something was wrong. Leaving the main party in base camp, Sher took off for Birethanti on foot. Actually he was running over the rugged terrain at a pace that would make Olympians proud. In fact, no professional runner could navigate the poorly defined and rocky trail with the speed at which the native Gurkhas did it.

As Sher ran, concern for both GB and the Dunlaps entered his mind. Living in Pokhara, the largest city in the Annapurna area, GB had become a trusted *friend over the course of many treks. Always friendly and smiling, GB was a favorite porter and could always be relied upon under the most stressful conditions.*

While he barely knew the Dunlaps, Sher was equally concerned for their safety. Tourism had taken a drastic downturn. While the incipient insurgency had raised concern abroad, it was the murder of several members of the Nepalese royal family that had focused international attention on the country's instability. On June 1, 2001, Crown Prince Dipendra, upset with family matters, had gone on a drug-induced rampage, shooting up the palace. King Dipendra, the cosmopolitan leader of the country, who had been educated at Eton, Harvard, and Tokyo universities, was shot with an AK-47. So, too, was Queen Aishwarya. In addition, six other family members lay dead. The twenty-nine-year-old Prince Dipendra then shot himself but did not die immediately. In an ironic quirk of fate the assassin became king, albeit for the briefest of time. He expired less than fifty-five hours later, and the former king's brother, Gyanendra, ascended to the throne. Throughout the constitutional monarchy there was great controversy concerning the new king's authority. Rumors ran rampant that he had actually engineered the events as a coup. As his rule was seen by many citizens as illegitimate, King Gyanendra's ability to command power was significantly diminished, leading to further destabilization. It was shortly after this turmoil the rebels broke their promise and struck.

Sher fully understood the implications of the kidnapping. If tourism had been off before, an incident in which Americans were killed would be the coup de grâce for his business. Falling off a mountain was an acceptable consequence for climbing the highest peaks in the world. Being abducted by terrorists was not. When in Kathmandu, Sher often watched the American television news networks. He was well aware of the Islamic terrorists, Abu Sayyaf and their exploits in the Philippines with the Burnhams. They were a missionary couple who were taken for ransom from a resort on the southern island of Basilan. As they were relatively poor people, there was no one to pay the fiddler. American policy was to not to accede to terrorists' demands for fear that would encourage further kidnappings. Held more than a year, there were repeated stories that aired about their fate. Unfortunately, Martin Burnham was killed in the battle to rescue him. Gracia also suffered a bullet wound but survived. These stories constantly reminded the viewing audience just how dangerous it was in the Philippines. Sher feared Nepal, too, would become known as a place to be avoided at all costs.

Birethanti was one of the few villages in the area that had continuous electricity. As Sher arrived he could hear a commotion emanating from a small plaza. Located just above the government's checkpoint where trekkers entering the Annapurna area were required to pay a tax, it served as a gathering place. There, on the stone pathway between the souvenir shops and restaurants, were the mayor and most of the local inhabitants. Before them on a blue plastic sheet was a headless body. One look at the red shirt provided all employees of Sherpa Enterprises, and Sher knew it was GB.

His concern that the Dunlaps had met a similar fate rose quickly as Sher realized he had forgotten to grab his cell phone when he left the camp. Though coverage was spotty, tourism demanded the ability to remain in contact. It was imperative that he get word to his company, as information of killed or kid-napped Americans would make headlines in the Kathmandu Press *and* The Rising Nepal. *Worse, the international press would have this all over the world in a matter of hours.*

While working late in his office, the American ambassador, a career State Department official with experience in counterterrorism, received the phone call. Having served previously in Nepal, he had been mildly alarmed, as the intensity of Maoist activities had grown. In fact, he recently had received a letter from a former Special Operations officer suggesting that insurgency in the Annapurna area was considerably higher than was being publicly acknowledged. The letter had also predicted the onset of kidnapping Westerners for ransom.*

Ambassador Colby was not surprised by the announcement. While he would make the obligatory statements of outrage and condolences, it had only been a matter of time before this occurred. An economic basket case, Nepal did what it could to take advantage of its natural beauty. The fees for permission to climb Everest ranged from $50,000 to $70,000 per person, and that didn't get the holder on foot on the mountain. It meant that people who came to climb or trek either had money or knew people who had money—at least that's how the Maoists would rationalize it.

"Get me Colonel Brooks," Colby yelled to his secretary. Like most military attachés, Colonel Brooks was an intelligence officer with an area specialty in southern Asia. He had also spent several tours in special operations and had close ties throughout the Special Operations Forces community. That would help with the coordination of this rescue mission.

For decades Nepal barely made a blip on the U.S. radar screen of national interests. Only after Secretary of State Colin Powell visited the country in December 2002 was any attention paid to the area. But even the new assistance

---

*While in Kanthmandu in December 2001 I talked with Ambassador Malinowski about this matter. In January 2002 I sent him the letter described in this vignette.

was hardly noticeable. Militarily it meant the United States provided money for a few weapons and conducted some training exercises in the mountains. These had been both conditioning and intelligence-gathering missions. As with many Special Operations exercises, the greatest achievement was in establishing person-to-person relationships. In Nepal, as in many other areas of the world, personal friendship and mutual respect between comrades-in-arms would be a critical factor in resolving difficult problems. Fortunately, there was a Special Forces team on such an exercise just south of the Everest base camp. With sweeping authority now granted to country teams, who were led by the ambassador, Colby could permit the SF team to conduct limited counterterrorist operations. Of course he would notify the President and the Secretaries of State and Defense of his actions. But Colby had the power to cut through bureaucracy and commit the military forces immediately. In similar situations the ability of on-site leaders to take action had received prior approval as part of the executive branch's initiatives taken in the War on Terror.

Within a day the Special Forces team could be relocated from the Everest Base Camp to the Annapurna area. However, there was some special equipment that would be needed for the operation. That would be flown in from the Pacific Command Special Operations Center in Hawaii. Prepackaged for deployment, advanced sensors, weapons, and environmental gear were ready almost immediately and designed to meet a wide range of contingencies.

In Birethanti the tinkling of a tin bell announced the arrival of the horse caravan. For literally millennia goods had been transported over the mountains by these surfooted animals bedecked in colorful harnesses. As the lead horse came into the plaza at around 10:00 A.M., the merchant called for the major, stating he had a message for him. Illiterate, as are about 65 percent of the people in the country, the merchant dutifully passed on a paper that he could not read. He had been ambushed while coming south on the Old Tibetan Highway and told to take the note to this village. The Maoists had confiscated a small amount of his goods, mostly food, as a form of tax. While he had not seen the main group, the merchant understood the note was important and that he was lucky to be delivering it alive.

Quickly the mayor called Kathmandu to alert the military to the message. It briefly stated that two Americans were being held for ransom and instructions would be provided shortly. If the military attempted to follow them, the hostages would lose their heads. Sher, reading the message, knew that the terrorists meant business. Any premature attempt at a rescue would end badly for everyone. However, he also knew that the longer they waited to take action, the more likely it was the trail would grow cold.

Born in a remote village, Sher possessed the innate tracking skills of the Pun tribe. These had been honed during his extensive military service, and he felt he could shadow the Maoists without being spotted. He also knew that while the terrorists could move swiftly across the rugged terrain, the relatively clumsy tourists would slow them down. For their sake, Sher prayed the Dunlaps could maintain a modest pace. Failure to do so, placing the terrorists further at risk, would mean certain death.

Knowing the Maoists would begin by following the zigzag but well-defined path for the initial hours, Sher decided to move in a straight line, thus reducing the distance he had to travel. Communication would be a problem. He told the mayor that he would use his now-retrieved cell phone from high elevations and would contact him with his location at least twice a day. That information was to be relayed to the military when they arrived. Sher was not yet aware that help would include American Special Forces supported by air capability usually unavailable in Nepal.*

Three days later the SF team, named Sigma for this operation, had arrived in Pokhara. They had been refitted with special sensors and sniper equipment and were prepared for deployment. For security purposes Team Sigma stayed at the Gurkha training camp on the north side of the town. With 350,000 residents, Pokhara was sure to contain many rebel sympathizers. While there would be some in the training camp, it was easier to control the access there. Also, helicopters frequented the area, so they would not raise as much suspicion as they would flying from a more remote site.

The members of Sigma were acclimatized to the altitude, making movement easier. However, no matter what their state of physical training, foreigners could never navigate the hillsides as fast as the natives. Therefore, Sigma was split into three subelements and assigned an equal number of experienced Gurkha soldiers, all of whom had served in the British regiments. There would be no language barrier during the operation.

Further, the Gurkhas had been carefully screened for the mission. Starting in April of 2002 the Maoists had stepped up their assaults. In major battles they had overrun military and police units. Although captives had been taken, none

---

*A major shift in Maoist warfighting capabilities was demonstrated on April 12, 2002, when they attacked military and police units, killing over 170 of them. Those captured alive were stripped naked, paraded through the village, and executed. On May 8 the Maoists again attacked, killing more than 100 soldiers.

survived the torture, humiliation, and public beheading that followed.* Each Gurkha attached to Sigma had one or more relatives who had been killed in those raids. This blended team represented an enviable match of capabilities. The Americans brought sensors, communications, and weapons. The Gurkhas had knowledge of the area and the enemy. In addition, their dedication was not a problem.

Beginning the first night, Sher called the mayor in Birethanti. While Sher had not spotted the Maoists, he had picked up their trail. During that transmission he was provided a new number to call. While it appeared to be the number of a Nepalese office, the phone rang in the U.S. Embassy in a special office manned by both U.S. and local intelligence agents. To ensure his phone batteries lasted Sher was advised to use only the assigned number. In addition to receiving Sher's information, the agents wanted to be able to determine his location.

After only two days a pattern of movement had emerged. Sher had sighted the Maoists and confirmed that the Dunlaps were alive. At first the group had attempted to put as much distance as they could between themselves and any trackers. Slowed by the limits of Gloria Dunlap's physical fitness, the terrorists moved in shorter bursts. After eighteen hours they took a sharp western turn. While a long ways off, the western area of the country was their stronghold.

In Kathmandu and Langley, Virginia, analysts began plotting the known locations. As the area had not been of strategic importance to the United States, only limited aerial photos were available. Therefore, a U-2 was deployed to make multiple surveillance runs over the area. Old but updated and versatile, the U-2s remained an important part of the photo collection tools. Flights were conducted at both midday and early evening. Fog was a serious problem in Nepal, often keeping small commercial planes grounded for hours. The day and night runs would provide updated photomaps and establish heat source patterns.

The overflights provided a snapshot in time. Needed was intelligence from staring systems that could monitor the area indefinitely on a real-time basis. To meet this requirement, aerostats were included in the deployment package from Hawaii. Based on photo analysis three sites were selected that would intercept the anticipated path of the terrorists.

The aerostats were equipped with both optical and infrared sensors. Tethered, these small aerostats could be floated at hundreds to thousands of feet. Their special coatings made them nearly invisible to the naked eye. While Nepal does not have an air force, they do use helicopters. To prevent accidental entangle-

---

*While the military in Nepal has a few helicopters, they do not have an air force.

ments, a no-fly zone was established for their pilots. They were not told that the aerostats had low-probability-of-intercept transponders that allowed the SOF helicopters to navigate the area safely.

The initial task was to establish the existing background. Based on the science of phenomenology, the analysts would be able to discriminate between the normal situation and external movement of the guerrillas. This was not phenomenology in the sense of ESP but rather an exact science based on mathematics.* The topography of the area would mask some terrain, but certain predictable areas would have to be traversed. Once the terrorists were spotted, teams would be sent in ahead of them. As even remote areas of Nepal are populated, care would have to be taken for insertions. Overt Nepalese forces were on the ground gathering intelligence from the local tribespeople. But that cut both ways. The Maoists often had their own people operating in these areas, and thus they would know the Army was chasing the guerrillas. There was inherent danger to civilians who were caught helping the government. Conversely, the soldiers often intimidated the civilians thought to have knowledge. Most of them were subsistence-level farmers, and this was a no-win situation for them.

After a week the Maoist band was spotted moving toward an area that was suitable for an ambush. They alternated staying in remote farmhouses and in dense terrain with cover afforded by bamboo thickets. The narrow mountain trails leading to their night encampments could be easily observed and defended by only a few guerrillas. Actually, they were doing a good job of varying their pattern, and getting close to them without being detected would be very tricky.

Sher, acting as a shadow, provided the ace-in-the-hole, but he was nearly out of energy for his phone. On the last transmission he was directed to move to an open location four hours' walk to the east. There he was picked up at 8:30 P.M. by the black MH-6 Little Bird helicopter that flew west from Pokhara, then dropped down and followed the Mayangdi river valley. With the baffled rotors and low-level flying at night it was unlikely that the pickup would be detected by any of the Maoist-sympathizing observers who seemed to be everywhere. Care was taken to avoid the suspension bridges that periodically allowed foot traffic across the waters that raged during the monsoon season.

The members of the 160th Regiment, known as the Night Stalkers, were extremely proficient at exfiltrations. Spotting the triangle comprised of three very small fires, the pilot of the Little Bird flared back and dropped deftly to the ground. Quickly Sher kicked the burning embers into the river, turned, and ran

---

*The U.S. Navy has the Strategic Phenomenology Section (formerly the Office of Strategic Phenomena) in the Naval Research Laboratory Space Science Division.

toward the helicopter. Without his knowing it, the thermal viewers grabbed a photo, which was fed into a tiny computer into which Sher's facial features had been entered from existing file photos. Receiving positive recognition, he was allowed to approach and board the craft. In less than a minute the Little Bird was darting back to the east and out of sight.

An hour later, back at the Gurkha camp in Pokhara the debriefing began. Sher provided a complete chronology of the trip from Birethanti. A skilled observer, he had mentally noted minute details about the guerrillas' movements and methods. Times, dates, locations, and tidbits about techniques were immediately transmitted back to the team of analysts waiting at Fort Bragg, North Carolina.

Although midnight in Pokhara, it was 1:15 P.M. at the modeling and simulation center.* The computer jocks had already loaded the existing digital terrain data into their model. That had been updated from the recent U-2 overflights. Area specialists assisted as the information Sher provided was inputted and the first runs made. While people on the ground in Nepal could conduct map reconnaissance, the advanced computing system would give an independent evaluation of probable courses of action by the Maoists. Within an hour the results were in. The most probable course of action was noted along with possible alternatives in relative rank order. The program also identified photographs from various databases that might be useful for mission planning. Another by-product was the development of maps designed specially for the area of interest. The scales of these maps varied based on the users' needs.

The impending action was now being integrated through multiple headquarters including Pacific Command (PACOM) on Oahu, Special Operations Command, near Tampa, and, of course, the Department of Defense in the Pentagon. However, Commander PACOM was the warfighter and the one whose feet would be held to the fire if things went wrong.

Via secure communications the senior SOF officer located in Pokhara was in constant contact with his SOF counterpart of Camp Smith near Honolulu. Since Army Special Forces unit would conduct the rescue mission, an Army SF lieutenant colonel had been sent to Nepal. First he had briefed Ambassador Colby in Kathmandu; then he set up a temporary command post with the team that was about to deploy.

Captain Andrews was the A-Team leader who would execute the operation. He, and most of his team, had been involved in chasing the al Qaeda fighters in the mountain caves of Afghanistan. That experience would come in handy.

---

*Nepal is five hours and forty-five minutes ahead of Greenwich Mean Time. It is a country that has decided to have its own time zone as a method of expressing independence.

Using both maps and aerial photos, the team searched for helicopter landing sites from which they could cut off the Maoists. However, as Nepal has a burgeoning population and limited land suitable for farming,* nearly every open area up to 10,000-feet elevation is under cultivation. Typically the farmers live on their fields or very close to them. Site selection was difficult. Since the insertion would include both U.S. and indigenous forces, there was no time to teach alternative techniques.

It was decided that the insertion of all three teams would be made at about 2:00 A.M. the following night. Under most circumstances this would have also been assigned to the Night Stalkers with their modified MH-60s. However, a single MH-53 Pave Low III helicopter of the U.S. Air Force 720th Special Tactics Group assigned to the 320th STS at Kadena Air Base, Japan, would be used for this operation. The Pave Low, designed for deep-penetration missions, was chosen because of the very advanced navigational equipment on board. They would be flying in the mountains at night, and heavy ground fog was very common. Sigma would be divided into three teams, each composed of four Americans and four Gurkhas. Sher would accompany the second team, the one most likely to make contact first.

Also assigned to that team was a canine handler with his dog, Rolf. Rolf was a Malinois, a Belgian breed favored by many police departments. Slightly smaller than German shepherds, Malinois dogs are phenomenally strong. They are known for their intelligence and controlled aggressiveness. As with similar breeds, their sense of smell is exceptional, making them great for tracking. Rolf was exercised daily and was trained to find a person and silently alert his handler. With a hand signal Rolf would bite without warning. Once he clamped his powerful jaws on a person only two things could make him release them—his trainer's command or Rolf's death. He would be useful in detecting sentries hidden along the trails they would travel.

Exactly at 2:00 A.M. Team Sigma Alpha was inserted into a small, isolated clearing. Ten minutes later Team Sigma Bravo successfully landed, followed by Team Sigma Charlie. The Pave Low then swung quickly to the west, made several fake insertions, and continued in a large loop to the north before heading back to Pokhara.

Once on the ground each team verified their position with GPS and prepared to move to their designated ambush position. In these rough mountains GPS was good for determining where you were but did not help much in navigation.

---

*With a rapidly growing population of 23 million, Nepal has an increase in population of 2.3 percent annually.

While the SF members used their night-vision goggles to see the trail, the Gurkhas found it easier to move based on their human senses. This was their terrain and no amount of training or technology could overcome the advantage that comes from having learned to survive with nature.

Team Bravo had about three miles to cover before daylight. There was a small shack hidden off the main trail at that location. If the projections were right, it would become the Maoists' camp the following night. Team Bravo wanted to arrive close to that location, establish observation points, and then hide before daylight. They hoped they could avoid contact with all local people.

Arriving near the target area before dawn, Team Bravo was surprised to find the shack already occupied. Through their thermal viewers it was determined that the people at the site were armed, but Team Bravo could not tell how many were present. Reporting back, they learned that the kidnappers were still a day's march away. This meant that Team Bravo had found the reinforcements and the ambush would be against a larger force than initially anticipated. While they were confident in their firepower, extricating the Dunlaps had just gotten a lot tougher. Thought was given to bypassing this site and attempt to intercept the band while they moved. Risks to the hostages and lack of preparation dictated against changing the operation.

The aerostats continued to provide information about the area. During the day it was determined that the kidnappers were heading to the shack. Orders were passed to Team Alpha and Team Charlie to begin moving toward that site as reinforcing elements. However, they were only to move when they could do so undetected. Again the skill of the Gurkhas would prove invaluable.

Team Bravo observed the shack from three locations. While the guerrillas did have guards posted, their demeanor indicated they felt quite secure. There were ten in this group, eight men and two women. Throughout the day they wandered about the area gathering wood and fetching water. Twice local farmers stopped by. It appeared that they brought food to the guerrillas. The team noted that when the farmers approached, several members of the group would remain hidden so the size of their force could not be determined.

At about 4:30 P.M. the kidnappers arrived in the area. Team Bravo saw that two members of the group with the Dunlaps went ahead and entered the camp. The rest, five plus the hostages, remained out of sight until summoned. As with Sher's earlier observations, clearly the group was well trained.

Near the back of the column but separated from each other were Jim and Gloria Dunlap. The team had photos to confirm identities. While Jim was easily recognizable, ten days had brought ruin upon Gloria. Hair bedraggled and clothing torn, she was limping quite badly. It was lucky that the Maoists had not decided she was more inconvenience than she was worth. Still, Team Bravo knew she would not be able to move quickly should the need arise. Once the ambush was initiated they would have to secure the Dunlaps and prepare to stand and fight until they could be picked up.

As darkness fell, Team Bravo decided to have a closer look at how the guerrillas were deployed. From a backpack came a micro-UAV. The nearly silent electric engine could take the craft right into the encampment. For twenty minutes the micro-UAV toured the site completely unnoticed. Pictures were sent back and the complete external layout was reconnoitered. However, the Dunlaps were moved inside the building and out of view. It was critical to know their exact position.

Captain Andrews sent back a burst transmission. All it said was 3:23. For the rest of the evening Team Bravo would quietly maneuver into position. A sniper and one Gurkha moved to the north across the clearing. At 700 meters, with clear fields of fire, targets would be a piece of cake. To ensure there were no ammunition-compatibility problems the Gurkhas had been trained with American M-16s. As these were lighter than the Gurkhas' typical Belgian FNs, they were delighted with the weapons.

For three hours the teams moved very slowly forward. Rolf led the way, scanning for unexpected sentries. A magnificent animal, he moved as silently as a shadow, periodically looking back for hand signals. At 150 meters from the shack Captain Andrews stopped. He signaled for Rolf to be withdrawn. The captain was about to use a sensor system that no one in Nepal had ever seen and he did not want to inadvertently upset the dog.

Once Rolf was out of sight, Captain Andrews reached into his pack and brought out two black rats, each outfitted with a small harness. In addition, they had tiny electrodes implanted in their heads. Well fed and cared for, RoboRats were about to make their first appearance on an actual battlefield. Quickly he tuned the cameras fastened to the harnesses. The picture wasn't great, but good enough to gain perspective inside the building. The first rat was released. Using a joystick, RoboRat One was steered toward the shack. Given its short legs, it took some time to arrive at the building and had no trouble entering. By directing the RoboRat along the walls the complete interior could be seen. The Dunlaps were huddled in one corner. Gloria looked as if she had already gone to sleep. Jim was awake but with a blank stare. Three guerrillas were inside eating. They barely paid attention to the hostages but were close enough that they could easily shoot them during the confrontation.

RoboRat One was recalled. The information was good enough that there was no need for more data at this time. Instead, Captain Andrews took off the camera and replaced it with a miniature munitions holder. Inventors had envisioned RoboRats as reconnaissance tools. They had not thought about them as weapons delivery systems. The trick would be to ensure they went to the right place at exactly the right time.

For the rest of the night Team Bravo, minus the sniper element, inched silently toward the objective. At 3:15 A.M. RoboRat Two was released. As the rat was much closer now, it would take a short time for it to make it to the building. Smelling food, RoboRat was tempted to move away. Short bursts to its brain kept it on course.

At 3:22 RoboRat Two ran across the floor near the now-sleeping guerrillas. As it went Captain Andrews expelled a number of flash-bangs. Very small, they wouldn't dazzle or deafen, but from a sound sleep they would be effective.

The captain received the word that the Pave Low was inbound, about ten minutes out. In front of him were two AH-6 Little Bird gunships of the Night Stalkers. They were ideal for this task. The AH-6 could hover in small clearings and deliver devastating fire. While the team might be able to take down all of the guerrillas, they did not want any surprises.

As they moved closer Rolf alerted twice. Both guards were dropped by a silenced .22 fired at less than five feet. At exactly 3:23 the flash-bangs detonated. Startled, the guerrillas jumped up. Most never made it more than a step before being struck down. The .300 Winchester dropped two men standing by the fire before they could reach their weapons. Farther out, two more lost their heads to the Gurkhas who preferred their khukuris *to the new rifles.*

Within ten seconds two team members were inside the shack and dispatched the three guards. They grabbed the Dunlaps and pushed them to the ground. Then with two Gurkhas they formed a shield around the Dunlaps to protect them from the firing that rang out. Care had been taken to identify each friendly soldier. They wore stripes that were visible in the night-vision goggles each pilot wore and could be seen by the sniper.

Realizing they were being overrun, several guerrillas attempted to make it to the woods. They were no match for the night-vision sights of the sniper. One by one he dropped them in their tracks. Suddenly the two Little Birds dropped into the clearing. There would be no more shooting tonight.

At 3:33 the Pave Low announced its arrival. The Little Birds moved to supporting positions looking for any residual fighters. By 3:35 Team Bravo and the Dunlaps had left the ground.

• • •

AMERICAN TOURISTS around the world are increasingly at risk from terrorists. Since this episode was drafted, climbers trekking to the Everest base camp in Nepal have been kidnapped and held for ransom. While they were able to negotiate their own release, others will not be as fortunate. It is known that al Qaeda has instructed its operatives to kidnap Westerners for both political and financial purposes.

Operations in extremely difficult terrain will require great skill and ingenuity on the part of our troops. The Himalayan Range is about as remote as it gets. In the previous scenario we see unique combinations of high technology and weapons centuries old. The use of rats equipped with cameras is only slightly beyond current state-of-the-art. However, the ability to obtain accurate information in time to react to it is critical. If you travel in out-of-the-way places, be thankful that there are Special Operations forces that can come to your rescue if necessary.

CHAPTER EIGHT

THE WAR WE WANT TO FIGHT

"MR. PRESIDENT, REGRETTABLY THE *time has come for us to address our mutual commitments and interests.*" *The voice of the Russian leader speaking in English was easily recognizable.* "*Geoffery, I am afraid we need your help. We had hoped we could keep the situation under control, but with this new problem . . .*"

"*Yes, Dimitri, we have been watching the buildup with great concern. I will get back to you within two hours,*" *he promised. The use of first names connoted familiarity from their prior meetings and a sense of personal urgency. The request was real, not pro forma.*

Key members of the National Security Council along with appropriate Cabinet members and the Director of Central Intelligence were already gathered. The problem was Russia's southern flank in the Transcaucasus region. For more than two decades the Russians had been involved in conflict there with no clear winners. The Chechen separatists still wanted an independent country. Russia claimed the guerrillas were terrorists, as evidenced by the constant interruption of oil flowing through the pipelines to export. The bitter fighting in Chechnya led to endemic atrocities on both sides. However, the oil resources of the region were too good to relinquish. This war and inability to stabilize their economy had drained the Russian resources to the breaking point.

To add urgency to these matters, most of those officials at the meeting had received personal phone calls expressing concern. The few who were not contacted directly had deputies who answered their phones. These calls were not from Russians but rather American oil executives. There was always speculation about how much big oil companies influenced foreign policy. This was a classic example in which these executives were willing to put troops' lives in jeopardy to protect the companies' income. Given their campaign contributions, they would not be ignored.

Moscow had formed an alliance with Azerbaijan and, with British and

131

American assistance, helped in the building of new oil pipelines from Baku via Georgia to the Turkish Mediterranean port of Ceyhan. A mutual defense pact had been signed at the insistence of the Russians. A fear by the people in Azerbaijan was that this was merely justification for military intervention should Moscow decide they wanted full control of the oil and gas reserves in the region. The United States had tacitly supported the alliance in an effort to keep the huge hydrocarbon wealth of the area in the hands of a modestly friendly country. Already the offshore drilling in the Caspian Sea was producing more than 100,000 barrels of oil per day, and this was just the tip of the untapped reserves. In addition, vast amounts of natural gas were trapped beneath the surface. Azerbaijan alone had proven reserves of 4.4 trillion cubic feet waiting to be exported.

Now Iran, bolstered by the forceful removal of archenemy Saddam Hussein and in concordance with Iraq's new government, was amassing armored forces just south of Azerbaijan. From every appearance it looked as if they would attempt to actively support their Muslim brethren, and this area had traditionally been a part of Persia. The ethnic ties between the people of Baku and Tabriz, Iran, were far stronger than those with Moscow. The real concern for the United States was the control of the oil fields and further concentration of economic power in an unfriendly state. If the Iranian army executed a quick dash they might be able to reach the refineries at Baku before the Russians could move their forces to the area. The strategy would be to grab the territory and sue for peace.

The Russians had the capability to stop Iran cold. However, that meant use of nuclear weapons before the forces got deep inside Azerbaijan. The White House understood this conundrum without an actual threat being made. It was declared in the national interests of the United States to intervene militarily. Picking up the phone, POTUS informed his friend that the United States could be counted on for armed assistance.

The new brigade level task force, Task Force Sabre, developed under the Future Combat System (FCS) had finished operational readiness testing and was fully deployable. As required, they could arrive in southern Russia within ninety-six hours. The warning order had already been given. Operation Southern Rumble was a go. The necessary airlift, predominantly C-17 Globemasters, C-141 Starlifters, and a few C-5 Galaxies, was being identified and moved into position near Fort Stewart, Georgia. Strategic airlift remained the long pole in the tent. Despite many years in which study after study demonstrated the shortfall, adequate funding for the big birds was still a problem.

Although there were armored forces stationed in Europe, the decision had been made to demonstrate the capabilities of the newly formed unit. While the decision to convert other brigades to FCS configuration had been made, success in battle would accelerate the funding from the Hill for the transformation. The Army could prove that the billions of dollars poured into transformation had been well spent.

As Iran had geared up, the NRO assigned more of its satellites to observe the area on a routine basis. Due to the potential consequences, one of the KH-13s was positioned to keep constant tabs on movements in southern Azerbaijan and northern Iran. The U.S. Air Force 19th Reconnaissance Squadron deployed with two Global Hawks forward to the base at Incirlik, Turkey, to be in direct support of the mission. Within twenty-four hours they would begin around-the-clock coverage of the entire region from 65,000 feet above. Between the NRO and the Air Force there would be constant day-night, all-weather surveillance of the assigned battle space until the requirement was withdrawn.

To get eyes on target as rapidly as possible, U.S. Special Forces soldiers were ferried in from Georgia. They had been on continuous training missions there since 2002 preparing the Georgian army to combat the guerrillas who lived in the mountains and periodically launched raids in southern Chechnya. Though scaled back, small teams were omnipresent for the purpose of maintaining re- lations with the people and providing continuous intelligence about what was transpiring. No satellite could sit in a bar and listen to the tone of the stories being told. These interpersonal relationships were invaluable. Now, however, the teams would be flown to Baku that night and be deployed close to the southern border the next day. While there might be some language issues for the SF soldiers, the new universal translating device developed by DARPA would be able to provide immediate capability to understand what was being said. There was moderate risk to these warriors, as continuous air cover was not yet available. Still, an SOF operator with a laser designator could do a lot of damage.

Task Force Sabre would be the key element in Southern Rumble, and it was unlike any of the predecessor brigades. At a flyaway weight of 6,000 tons, it met the Army's FCS goal of one-eighth that of an old mechanized infantry brigade. The unit was highly mobile yet extensively armed with state-of-the-art sensors and light vehicles—including many robots. Nothing topped twenty tons, and there were only sixteen vehicles that heavy. That was a far cry from the old M1-A1 Abrams tank that grossed out at as much as seventy tons. Any vehicle in TF Sabre could be carried by the venerable C-130 Hercules aircraft.

Though light, TF Sabre packed far more punch than its older counterparts. Comprised of complex systems, the organization had been built from the ground up. Sensors, shooters, and support vehicles were all in communication with a *command and control system both internally and externally compatible. Not only were all elements of the brigade netted together, but also their communications systems could talk directly to the Air Force and higher headquarters. Such complete integration had been too long in coming.*

Manning TF Sabre were soldiers with equipment designed under the Objective Force Warrior program. Highly skilled, these soldiers received constant up-to-date information delivered to each one through heads-up displays and audio cueing. While every soldier had a personal weapon, most could direct firepower from heavy systems located several kilometers away. Through encrypted GPS transmissions, commanders at all levels knew the exact location of each soldier. Biosensors constantly monitored and reported the physical condition of each individual. Though well protected, should they receive wounds their uniform could initiate lifesaving measures such as applying a tourniquet. This was the epitome of Army transformation.

Within the allotted ninety-six hours from alert notification the entire task force had arrived in Baku. Immediately they began moving south toward the border. Their high-speed vehicles traveled at rates faster than most private cars. Cross-country, their mobility would make off-road racing enthusiasts drool. Speed and maneuverability were key to their survival in a high-intensity battle. These ground vehicles incorporated signal suppression technologies previously reserved for stealthy aircraft. Special materials reduced the heat emanating from the engines while acoustic baffling stifled the sound. Though these vehicles had adequate ground clearance, they were not very high and unique paint reduced the radar reflectivity.

During the movement Colonel Heal received an update from national reconnaissance indicating that the Iranians were now holding their troops about ten kilometers south of Azerbaijan. Obviously they knew that an American unit was on the ground. That would mean the U.S. Air Force and Navy would probably establish air superiority but not air supremacy. The difference meant that for brief periods of time the Iranians could fly in carefully selected areas.

The Americans expected the Iranian forces to move quickly, but they waited. It is when a unit is arriving and getting organized that it is most vulnerable. Actually, the delay was perceived as beneficial to both sides. The Americans were able to deploy their battalions without excessive pressure. The Iranians wanted to take away the known U.S. airborne sensor advantage. Rainstorms were due at this time of the year and cloud cover quite pervasive. It was believed that the tanks could accomplish their mission to seize a large amount of territory before the Americans could capitalize on superior airpower. Besides, Russian intelligence sources reported that only a light brigade had arrived. Certainly four armored divisions could overwhelm them before the Americans could bring in their death-dealing Abrams tanks. Since Desert Storm no adversary wanted to go nose-to-nose with those behemoths. Further, as Azerbaijan is very predominantly of the Muslim faith, these outsiders would be resented and a significant guerrilla movement was expected to inhibit U.S. forces.

The Air Force weather squadron was the first to observe the changing patterns in the upper atmosphere. Though predictable, the observation gave U.S. commanders and their Russian counterparts the head start in planning the next phases of the mission. The operational concept was simple: See first, shoot first, kill first. If the heavier Iranian tanks got too close it could be difficult to stop them. Therefore, the attrition had to begin as soon as they crossed the border— even before intent to attack was proven.

The Reconnaissance, Surveillance, and Target Acquisition (RSTA) Combat Command went into high gear. This was a new form of intelligence-gathering battalion. They brought an innovative set of air and ground sensors. Included were smaller tactical UAVs to augment the Predators and Global Hawks already operational in the area. The RSTA battalion also had new mast-mounted and tethered sensors that could see deep into aggressors' rear areas. Most important, these sensors could distribute their data in real time to every echelon that needed it and were netted to fire control systems.

Splitting into small elements, the battalion raced as far forward as practical. Iran would employ heavy concentrations of artillery fire before they moved forward, and the RSTA vehicles could not withstand being hit. Therefore, they would have to remain out of range of the big guns yet close enough to determine how the attack was developing.

The most dangerous approaches were relegated to remote-controlled robotic sensors. Small and highly mobile, they could hide unobserved and report back both optical and infrared signatures. Their miniaturization and redundancy added to their survivability. Losing robots was not like losing soldiers.

Elements of the maneuver combat battalions were also busy near the border. Remaining beyond line of sight, they implanted antitank minefields along all of the high-speed avenues of approach. The front could not be covered in its entirety, but these minefields could take away use of the road network. The mines would slow the advances and force the Iranians to either clear the area or go around it. Of course suicide mine clearing was a known tactic. During the war with Iraq, Iran would have volunteers wearing religious headbands tromp through known minefields. Those who were lucky enough to detonate a mine were promised immediate martyrdom. Many young zealots made the trip. To set off antitank mines the Iranians would probably zoom in with old SUVs and buses.

What they hadn't counted on was smart mines with magnetic sensors that could discriminate between armor and civilian vehicles.

Reports from the SOF elements were forwarded to the commanders. As expected, several villages were preparing to actively support the invading forces. The locations of weapons caches and identities of key personnel were passed along. As part of the joint U.S.-Russian effort, members of the local intelligence units had been screened using LEADS deception detection technology. A number of agents with questionable loyalty were quietly removed from service. The trustworthy ones were provided with the identities of the suspected collaborators. As soon as any shooting started, the wanna-be guerrillas would be immediately apprehended. If lucky, they would be held as prisoners. Most would not be lucky, but this area of the world plays by its own rules.

Around midnight on Thursday the Global Hawk reported a high concentration of new heat signatures on the Iranian side of the border. Obviously, the war was about to get under way. Intelligence placed the size of the unit at somewhat over two divisions, with two more ready to reinforce. The formidable TF Sabre vigilantly waited for their first engagement since being formed, a test of technology and training that would set the course for the whole Army to follow.

The rains began as predicted. The attack would come before the ground became too muddy and began to inhibit the combat effectiveness of the armor. Speed and shock were what they counted on. Even absorbing some punishment, they anticipated to be at the outskirts of Baku in two days. The rest of the action would be diplomatic.

The report from the orbiting JSTARS stated their moving target indicators showed at least twenty-two armored and mechanized battalions running straight toward the border. First crossing would take place within thirty minutes. Fifteen minutes later, forward positions began reporting heavy artillery fire. The Iranians seemed to be prepping the general area where they would cross, as opposed to shooting at specified targets. Though too far out for long-range ground weapons, the counterbattery sensors on board the robots began reporting azimuth and range to the guns. These readings were sent back to the centralized command and control center to be passed to air units that were now en route.

The first shot by American forces came from a forward-deployed robot equipped with a compact antitank missile. It had been hidden less than 300 meters inside the border and tied into an optical sensor. Despite advances in technology, politicians were still not sanguine about a robot that could sense and kill without a human making the decision. It had been a remote operator who had identified the target as an Iranian tank and begun the assault sequence.

Once permission to fire had been transmitted to the robot, it took over the process and located the most vulnerable part of the vehicle, then fired. Once the external fuel tank was ignited, the armored vehicle instantly became a blazing inferno. Uncharacteristically, the first casualties went to the attacker.

The JSTARS sensors depicted two main thrusts about twenty kilometers apart. They directed the first bombers arriving on station to positions directly above the massed armored forces. The wave of bombs released was not the typical precision-guided munitions. Rather, there was a release of submunitions that parachuted down through the clouds. Each of these had sensor systems that could locate the heat from an engine and align the attack. An explosion—relatively small by antitank standards—would follow and the warhead would form itself to penetrate the top of the tank.

The top-attack technology, long under way, had been embraced more fully after a review of Desert Storm destruction. That study determined that more than 90 percent of the targets destroyed would not take more than a five pound warhead. Why waste a big missile against the frontal glacis when a small bullet in the right place will do? These small but expensive submunitions were very effective when armored forces were moving close together in the open.

Another feature had been added. If the submunitions did not spot a target during descent, it would land, automatically right itself, and become an antitank mine. The sensor continued to search for targets until the battery gave out or an encrypted remote control signal placed it into a dormant mode.

Despite the unexpectedly high initial casualties, the Iranian armor pressed forward. As they did so both JSTARS and the RSTA sensors began isolating targets. The fire control center began assigning fire missions. Priority went to the quickly deployable NetFires missiles. The Army had pushed for longer-range weapons by which they could shape the battlefield. NetFires provided that capability with missiles that could fly out in excess of fifty kilometers and find and attack specified targets. The NetFires variant being used in the initial phase was the loiter attack munition or LAM. After the vertical launch it would dash to a specified area. Then the LAM could loiter above that location, flying in circles for up to sixty minutes seeking armored vehicles with the onboard LADAR and automatic target recognition sensors.

The cueing robots of the RSTA battalion were an ideal compliment. They would locate moving armored targets and relay range and azimuth data. Thanks to GPS the exact location of the robotic sensor was known. Further, because they operated in netted teams, the fire control center could isolate and designate targets without fear of redundancy. The NetFires missiles were relatively expensive, and optimally there would be one kill for each missile. While not practical, they did have about a 90 percent kill ratio. This combination of netting fires and use of robots for targeting circumvented one of the age-old problems in armored warfare. How can a commander be sure when an enemy tank is destroyed or at least not a threat? In the past each wave of tanks would put extra rounds into an enemy vehicle just to be on the safe side.

Unfortunately, in modern warfare extra shooting both is unacceptably expensive and overburdens the logistics system. Even with advances in automated resupply, the system will struggle to keep up with any high-intensity battle. However, NetFires was easily supportable. An entire 2,500-pound package could be delivered to any location. That container held fifteen missiles and a fire control system. Modular in nature, it could be mounted on an HMMWV or ground-deployed in a static situation. Ideal for this battle, NetFires was placed sufficiently far to the rear that artillery would not be a threat for quite a while.

The Iranians had expected our airpower to dominate the airspace but were surprised at the effectiveness of the all-weather bombing. They responded by launching a rudimentary air attack of their own. The purpose was not to defeat the U.S. Air Force but to detract from their ability to bomb accurately. For a few hours this tactic worked, as F-22 Raptors and F-16 Falcons were called upon to engage the Iranian fighters, which were mostly Chinese-made but highly maneuverable SU-27s. The stealthy Raptors, armed with eight air-to-air missiles, were devastating. However bravely they fought, the Iranians were no match for the advanced U.S. fighters, and soon the skies were clear of adversaries. To ensure there was no further interference and to punish Iran for initiating the war, the airfields supporting the attack were destroyed in a massive B-2 strike. Also hit were key refineries, to inflict limited economic damage.

On the ground there was considerable shooting and the Iranian forces were reporting a substantial number of kills. However, they still did not comprehend that what they had engaged was only the vanguard of attack robots. To troops inexperienced in robotic warfare these robots seemed just like traditional forces. In reality, not a single American had been lost to enemy fire. The seven losses that had occurred were all due to accidents, as even remote combat is inherently dangerous.

The armored attack was also being befuddled by the ability of the attack robots to relocate when fired upon. Being relatively small, they were difficult targets that could hide behind walls and in brush. These intelligent robots also worked as teams. When one was attacked, others would locate the assailant and maneuver against it. The command to fire was still executed by a human. However, at this stage of the conflict the operators were giving permission to shoot anything that moved. This far to the front, identification, friend, foe, or neutral (IFFN) was not a major concern. Should the fighting get close, avoidance of fratricide would become paramount.

Willing to take heavy casualties, the Iranian armor pressed onward. The forward-deployed RSTA and maneuver battalions were now directly engaged in the battle. The rain had increased in intensity, and visibility of optical sensors decreased. However, these conditions allowed the small mobile robots to get quite close to the attackers without being detected. The high winds and poor visibility did prevent the use of the light tactical UAVs for target designation. Still, the combat unit's beyond-the-line-of-sight short-range missiles took their toll. Soon the first wave was stalling a good seventy kilometers from their objective.

That was the good news. The other news was that the second wave of Iranian armor was now racing through the area that contained the smoldering tanks of their comrades. By now most of the robotic mines had been detonated and many of the RSTA sensor robots damaged or destroyed. The remnants of the mobile minefield did attrit a few tanks, but not enough to even cause the divisions to slow down.

More NetFires LAMs were dispatched, but without the benefit of the full complement of netted ground sensors as a cueing mechanism. Still, most would find a target to attack and subsequently destroy. Above, JSTARS continued to relay critical information to the ground commanders. This allowed repositioning of forces to intercept the main thrust. In addition, because of comprehensive IFFN systems on the FCS vehicles, warning was given to units that might otherwise have been surrounded. The ground units did not want to engage in eyeball-to-eyeball fights if avoidable. The mission was to hit and run before they became decisively engaged, and the improved mobility allowed for that.

Of real concern was the Iranian artillery that was now advancing behind the armored units. If they got within range of TF Sabre units they could inflict serious casualties. While TF Sabre's vehicle did have advance ceramic armor protection, they could not withstand direct bombardment from heavy artillery. Survivability came from speed, maneuverability, and remaining hidden while striking first.

The U.S. combat maneuver battalions began striking tanks with their shorter-range indirect fire missiles. In 2002 the Crusader artillery program had been canceled, much to the chagrin of its Army supporters. With the extended ranges and rapid rate of fire capability it would have been very useful in this battle. Of course it did not meet the low weight requirements for the FCS. In the perennial battle that pits funding for guns against missiles, the politically savvy Huntsville-based command had again come out on top, and missiles dominated the brigade's weapons systems.

From above, the unique signature patterns of artillery batteries were easily discernible. These were immediately targeted for an area attack. In the 1990s the U.S. Army developed SADARM, which stood for Sense and Destroy Armor. These were top attack submunitions, two of which fit into a 155mm howitzer shell. The initial concept called for a heavy artillery unit to fire a number of these above an armored unit.

Each SADARM submunition would seek a target and fire. However, the old artillery was far too heavy for the FCS brigade. Therefore, an air-launched cruise missile (ALCM) variant was developed for the Air Force. This provided long standoff distance for the aircraft, and each missile could destroy one or more armored companies or artillery batteries with a single missile. Only when ground targets were massed was this a cost-effective weapon.

Deprived of artillery and having sustained casualties in excess of 60 percent of all of their armored vehicles in less than twenty-six hours, the Iranian corps commander radioed a request for permission for his forces to withdraw. Instead of suing for peace on their terms, it seemed more likely that additional forces would be needed to defend Iran's border areas with Azerbaijan.

Tehran had established a regional headquarters near the northern city of Tabriz. As a permanent site it was well fortified with extensive underground facilities. During the course of the battle Rivet Joint, the U.S. Air Force RC-135 electronic surveillance planes had been monitoring this site's radio communications. While jamming could have been successful, it was determined that the intelligence value coming from the transmissions was greater than the tactical advantage of taking them off the air. Now it didn't matter anymore.

The decision to attack the headquarters at Tabriz was made with concurrence from the National Command Authority. The weapons would be three BLU-118Ds, thermobaric bombs that were greatly improved from those used to destroy the caves in Afghanistan. The ferocity of the fuel air blast, capable of collapsing the buried compartments, would send a clear message to everyone in the area.

The raid began with F-117 stealth aircraft identifying and eliminating the Iranian air defense systems in the area. Since these locations were known from prior reconnaissance, it did not matter whether their acquisition radars were turned on. For those that did light up the sky, the second wave of fighter-bombers released their advanced radiation-seeking missiles that tracked the signals back to their source.

Twenty miles away, people in this earthquake-prone region ran from their homes. The shock seemed like a jolt greater than 8 on the Richter scale. Some, seeing the fireball rising from the new crater, believed that the United States *had dropped a nuclear bomb. In midsentence the radio went dead, and the corps commander never would hear the answer to his request. It was really moot anyway. Survivors were already retreating without orders. The constant onslaught of precision missiles had taken a toll that no analyst expected. If Iraqi Freedom had set the mark for devastation with modern weapons, Southern Rumble had raised the bar dramatically.*

The counterattack came swiftly. A battalion of the newly acquired Comanche helicopters had reinforced TF Sabre. Long delayed and frequently restructured, the program to put Comanche on the battlefield had been an arduous one. An excellent tank killer, the RAH-66 would prove the cost had been worth it. Unlike previous helicopters, Comanche could function as both a reconnaissance and attack craft. It could also integrate input from UAVs searching for targets. With an eight-kilometer range, the Longbow Hellfire missiles could engage a tank before it was aware these helicopters had spotted it. With a dash speed of up to 175 knots, no ground vehicle was going to be able to get away. For armored personnel carriers, the 20mm three-barrel Gatling gun would rip them to shreds.

The highly coordinated combined arms assault sped across the battlefield at an unprecedented pace. Occasionally troops would dismount and over-watch an enemy unit that had been bypassed. Once signaled, they would usually surrender rather than dying with their tanks. The tactical UAVs relayed targeting information and precision guided missiles would be launched on the move.

The Special Forces soldiers had never left the area. As the Iranians approached, they merely went into prepared redoubts and hunkered down. From these observation posts they had relayed critical information about troop movements. Retreating in disarray, the Iranian vehicles became targets of opportunity. Armed with laser designators, the Green Berets would illuminate a target just long enough for a missile to acquire it. Seconds later there was a burning hull.

In six hours the borders had been restored. Following explicit orders, the U.S. forces did not pursue the disheveled adversary onto Iranian soil. Any further action would be left to the Russians.

Even before the few survivors made it across the border, Tehran was signaling to both Washington and Moscow that they wanted to halt the action. For the United States, the petroleum supplies were safe and Russia owed us a major favor. For the next decade there would be no threats of reduction in oil production for the sake of manipulating the price. As for Iran, Russia worked out a separate deal. It included loss of some of the oil-rich fields near the border in the Caspian Sea.

• • •

THE PRECEDING scenario is important as it portrays the thinking of much of the mainstream military. They want Desert Storm Redux, in which the enemy stays in the open and fights a conventional battle. Using robots and long-range sensors and standoff weapons, they hope for a pristine engagement sans close combat. The result, they hope, is the ability to destroy the adversary with minimal casualties to U.S. forces.

This engagement was based on the requirements put forth by the Army for their much-vaunted Future Combat System (FCS). The FCS is to quickly deploy to any location and be prepared to fight independently for several days. In the end, times and loads will prove unrealistic. Not accounted for was the insertion of forces into a hostile environment. Robotic platforms will increase combat capability, but the Army will have more difficulty integrating them than it anticipates. However, if any adversary does oppose our forces in a manner similar to this scenario, they will be completely destroyed. Of course, they already know that.

CHAPTER NINE

THE WAR WE WILL FIGHT

THE AXIS OF EVIL *shuddered, convulsed, and imploded. Through acquiescence to the demands of the increasing radical mullahs, the House of Saud tried to hold on to power. Eventually, with population growth, the religious base shifted and with it political control. Sensing disquietude and in order to placate their citizens, the princes of Saudi Arabia took steps that continually alienated the American people. Despite private assurances between governmental officials, the U.S. media lambasted these actions, and open support, including military assistance, became politically untenable for any member of Congress. When it came, the near collapse of the kingdom surprised no one.*

The struggling government of Afghanistan, an accumulation of disparate entities agglomerated by America never did have a chance, even with considerable foreign aid. So, too, had fallen Hosny Mubarak in Egypt and Gen. Pervis Musharraf in Pakistan. With them went all vestiges of moderation toward Western culture. In each of these cases, attempting intercultural balance had lost to theocratic hegemony. Not that moderate voices were silent; they just lost out in the cacophony invoked by zealous religious leaders of the undereducated masses.

Europe and Russia experienced internal strife from pressure exerted by fundamentalist Muslims who had arrived during the mass migrations of the 1990s. Shortly after the elections in November 2002 Turkey became far more conservative. Though not supported by Turkey's military, the American forces were asked to leave their territory. The War on Terror devolved into a series of skirmishes conducted under the aegis of independent alliances. Massive military resources were consumed in waging minor conflicts and maintaining troop commitments to hold tenuous peace initiatives together. Stretched thin, albeit against the wishes of those in governance, the United States needed to choose wisely the battles in which it would engage.

Recognizing the overextended position of American military forces and the

143

economic encumbrance of current conflicts, it was determined that an Islamic coalition could safely increase pressure on Israel. That resulted in Hezbollah and Hamas becoming far more active, with the renewal of suicide bombings on a regular basis. They also periodically fired rockets from southern Lebanon toward Israeli settlements.

The Israeli intelligence service identified that extensive external support was being provided to the terrorist organizations. Despite retaliatory air attacks, the bombings continued and even increased. Finally exasperated with the unwillingness of Arab governments to put a stop to these activities, Israel once again invaded Lebanon. Instead of the anticipated limited strike at the refugee camps, the Israeli Defense Forces moved quickly all the way into Beirut, just as they had done in 1982. There was no stopping the Israeli armored forces supported by helicopter gunships and fighter-bombers.

Tacit approval for the plan had been given by the U.S. President. Israel did not need such approval, but it ensured support if things got out of hand. Publicly U.S. government officials claimed that this invasion might be too extreme a measure. However, the mood in this country had changed. Not only had American soldiers died in various battles with Islamic militants around the world, but also terrorists had repeatedly attacked U.S. citizens traveling abroad. Finally, the attacks on the mainland had nailed the case. Several suicide bombers had successfully blown themselves up in shopping malls across the country. A chemical incident had occurred at a large rock concert, killing more than 300 and injuring nearly a thousand more. America was reaching the same pain threshold as Israel and was not about to hold them back. Orders went out to dispatch a second carrier group to the Mediterranean and to alert the air bases in Italy.

All of the Islamic governments in the Middle East were upset with the magnitude of the Israeli response. It had been understood that there would be air strikes and sanctions, but a full invasion of Lebanon was another story.

Three weeks later the response came. Syria, backed by Iraq, entered the fray, sending armored units from Damascus all the way to the Mediterranean. Even normally moderate Jordan voiced support for the operation while Saudi Arabia pledged troops and aircraft. Egypt established a major buildup of forces along the Suez Canal and threatened to invade across the Sinai. The small peacekeeping force in the area was totally overwhelmed with these actions. The U.S. battalion that had been on duty in the area since 1981 as part of the multinational mission was given orders to pull back toward Israel. Thus from the beginning it was clear that we would not remain neutral.

Two divisions of the IDF were now cut off from resupply. In Beirut itself the fighting could be problematic. Well-equipped militia units had entered the city as soon as the Israeli objective became clear, and more were coming. Hundreds of thousands of civilians were trapped in their homes. However, they were very sympathetic to the goals of the Islamic freedom fighters and, if necessary, would die to support their cause. Beirut, the city that would not die, was again about to be ravaged.

The recently rebuilt and modernized section of town on the west end of the peninsula consisted of densely constructed high-rise buildings—the legendary urban canyons most feared by military planners. Fields of fire, except along a few relatively straight streets, were often less than 100 meters. The multilevel buildings allowed antiarmor snipers to shoot down, while the tanks below could not elevate their guns high enough to return fire. Wisely they pulled south to the international airport, from which they could dominate any local battle. If they moved farther south, the IDF forces could smash through the Syrian lines but in so doing would relinquish their objective.

Israel now faced a multifront war with adversaries to the north, east, and south. Therefore, they could not send another large force to reopen the routes through Lebanon. Since inception they had considered the possibility of national extinction. If it came, it would not be without a fight of biblical proportions. Talks between Washington and Tel Aviv were extremely frank. If necessary, Israel would use its nuclear arsenal to regain freedom of movement for the IDF. Initially enemy armor units would be destroyed; if this was not convincing, Damascus would become a footnote in history. That would undoubtedly prompt a massive chemical or biological attack against the citizens of Israel.

As this sector of the world is among the most heavily monitored, our satellites confirmed the situation as described. The President told the National Security Council he did not have Armageddon as a political agenda item. The initiation of a nuclear confrontation in the area would have disastrous consequences throughout the world. Intervention had to be made—and fast. The U.S. Sixth Fleet, guardian of the Med. since the early nineteenth century, was already on full war alert. Elements of the II Marine Expeditionary Force, supported by CTF 61, the Amphibious Ready Group afloat with them, would have to make a forced entry over the shore. An Army FCS brigade would deploy from Germany to assist in establishing gaining control of the city and reestablishing the lines of communication for the IDF forces.

The first Naval presence would be from CTF 69. The recently commissioned USS Texas, SSN 775, a Virginia class submarine, was patrolling just off the coast of Lebanon when the invasion began. With SEAWOLF quality stealth capabilities and advanced sensors and strike munitions it could function in the littorals quite safely. However, since they wanted a peek at the port area, it was decided to dispatch a UUV. Such UUVs were designed to operate in mined areas and shallow areas not hospitable to the larger submarines. Through the UUV's enhanced sensors, accurate imaging of the critical areas of coastline could be reconnoitered.

With the IDF holding the airport, some forces could be flown directly in. However, high-rise buildings to the north would allow snipers to fire directly at planes and dismounted troops. Also, there were a million places from which observers might call in artillery fire from the guns hidden in the Lebanon Mountains, which ran east of the coastal plain. Great care would be necessary if an air assault on the airport took place.

To put pressure on the northwestern part of the city, one Marine Expeditionary Brigade (MEB) would assault and capture the port. As it dated back to the Phoenicians in 1500 B.C. the port had endured several such attacks. The second MEB would come ashore at the beaches near the airport and assist in securing the runways so that the U.S. Air Force could bring in the Army FCS brigade. Once that was accomplished, these units would begin to clear the city and eliminate the militia forces.

While the Marine Corps routinely trains for and engages in assault landings, the Army has barely acknowledged the problems associated with forced entry, especially when in close proximity to the enemy. This would not be a simple raid that could be executed by a lone Ranger battalion. The Rangers were good but not sufficiently armed or protected to take on an entrenched adversary with heavy artillery and armor.

The first unit on the ground was a combination of Special Forces and Air Commandos. Flying from Aviano Air Force Base on the night that the orders had been issued, these Special Operations forces had conducted a high-altitude/low opening (HALO) parachute jump onto the international airport. These skilled parachutists were able to exit the plane well out over the Mediterranean and guild to the rendezvous site that was illuminated with an infrared marker. All twenty jumpers landed within a ten-meter radius and were immediately greeted by IDF soldiers, who were expecting them. The Special Operations task would be to coordinate the coming assault and provide direct communications with the U.S. Air Force and Navy bombers that were already sweeping the area. As had been eloquently demonstrated in Enduring Freedom, the combination of trained operators on the ground for target designation and precision-guided munitions was hard to beat. Unlike in Afghanistan, these troops could not move about the countryside for better target acquisition. A few brave Israeli soldiers had ventured out in attempts to locate the militia's observers. Only two came back, and they were in hideous little pieces.

The arrival of U.S. airpower took a great deal of pressure off of the Israeli Air Force, which was now fully committed in all directions. The intentions of the United States were made clear. As long as major battles did not erupt, our support would be constrained to the Beirut area. Of course that included suppressing the artillery that continued to snipe at the airport on a regular basis.

The night after the Special Forces team had landed, Global Hawk began routine surveillance. Not only did it cover Beirut; it could observe movements in the Bekáa Valley to the east, which was being used as a staging area for both Syrian and militia units.

In the days and nights that followed, SEAL teams patrolled the beaches at both proposed landing areas. Dropped well out at sea, the new Advanced SEAL Delivery System, a minisubmarine that was hosted by the Texas, allowed them to stay dry while maneuvering to a site close to their objective area. The ASDS had significantly improved the SEALs' ability to operate for long periods of time. In the earlier model the SEALs were wet from the beginning of the mission until the end. Despite excellent wet suits, and equipment, maintaining core body temperature was a serious factor and of limited effectiveness. As any scuba diver knows, even the relatively warm waters of the Mediterranean would lower one's temperature and reduce the ability to function.

The coastal waters were searched for mines, but none were found. After a week, SEALs were able to enter the harbor and maneuver on land. These missions were kept to a minimum for fear that one or more SEALs might be captured. While none of the SEALs was privy to the invasion plan, their presence would be taken as an indication of our intent and there would be great political value in holding American prisoners.

D-day. At 0230 hours the C-141s made their first pass from south to north, then banking quickly to the west on out to sea. At 600 feet, it took less than a minute for the men of the 2d Battalion, 75th Rangers, to be on the ground. Within ten minutes all of the sorties had disgorged their loads and headed back to Italy. There had been no unusual activities to alert Beirut to the beginning of the U.S. operation. However, as soon as the planes could be heard overhead, inaccurate sniper fire began pouring in from the north and east. It was only a few minutes later that the airport started receiving fire from the howitzers in the hills.

Already circling overhead were two converted 757 aircraft now employed in the Big Gun concept. Civilian aircraft from the Civil Reserve Air Fleet, or CRAF, program were used instead of Air Force assets. That was because of the shortage of lift aircraft in the inventory. Also, they could carry more ordnance than traditional bombers and had long loiter capability. Loaded with retrofitted bombs, they waited for targets of opportunity, expecting the artillery to commence firing during the American air assault. Because of the heavy concentration of non-combatants, the rules of engagement stated that only confirmed hostile fire could be engaged. The area had been segregated into quadrants permitting an ease of restrictions in the mountains.

The aircraft flew quite high. Although there were no fixed air defense sites around Beirut, many man-portable Chinese-made HN-5 ground-to-air missiles had been brought into the area. These were a variant of the old Soviet SA-7 Grail missiles that could reach to just under 15,000 feet altitude. Any militia member with one could climb onto a rooftop and take a shot, so caution was advised. In fact, the exhaust plumes from the solid fuel rocket engines of several missiles could be seen as they ineffectively spewed forth into the night sky. Of course a huge volume of small-arms fire accompanied the rockets. The aircraft were well out of reach, but in much of the Third World shooting into the air is a common practice. They have never yet figured the simple law of physics that states what goes up must come down!

As targets were identified the bombs were simply pushed out of the plane. Gravity brought them down and their guidance collars directed them to their assigned targets. Unlike traditional artillery units, these guns were deployed individually or at best in pairs. Well hidden and sometimes rolled back inside buildings after firing, they did not provide the concentration of targets associated with an artillery battery. Each had to be located and destroyed in turn. It also seemed that the gunners had no centralized command and control system. When they felt like taking a few shots, they did. The tactic lacked the effectiveness of concentrated fires but also created a randomness that raised tensions of the troops hunkered down at the airport.

At the Beirut port, SEAL teams had already infiltrated the area. There were no beach defenses per se, but small units were running around independently. As soon as they heard the advancing roar of the CH-53 Sea Stallions they instinctively began to take up firing positions. As targets came into view, the SEALs opened fire. Within a minute they would be heavily reinforced and the secret was out. Ahead of the CH-53s came flights of AH-1Ws, Super Cobras. They strafed the buildings immediately surrounding the docks. No massive bombardment by the impressive naval guns was allowed due to the close proximity of large numbers of noncombatants. However, at this hour of the night it was assumed that civilians would not be congregating on the docks.

Quickly the Sea Stallions disgorged their loads, typically about thirty-seven troops. The intent was to establish a foothold at the port and drive south into Beirut. Heavy fire was exchanged between the troops and small militia units *located at many points around the base. In addition, the Super Cobras encountered automatic weapons fire from the roofs of the tall buildings to the northeast and south. Several were hit, as were the Sea Stallions.*

While the fast-movers could fly above the range of the HN-5 air defense missiles, the helicopters provided excellent targets. Without command, the militia seemed to target the gunships first. Three went down in the area surrounding the port while five others limped back to sea and the relative safety of the waiting Tarawa-class LHA (Landing Helicopter, Assault) ship. Unlike in the well-known story of Black Hawk Down, there would be no land rescue party coming to get the crews. The fight was on in earnest. Even though the aircraft had gone down within a kilometer of the assault force, the ferocity of fighting would prevent infantry units from joining them. Also, the high buildings prevented other helicopters from approaching the area. The best that could be done was to provide covering fire and prayer.

Behind the helicopter assault came a wave of LCACs (Landing Craft Air Cushioned). Their ability to travel at up to forty knots had been a major innovation in seaborne invasions. The old craft maxed out at eight knots and were very vulnerable as they approached the shore. The LCACs could actually drive up onto the shore and drop their loads.

The initial assault force, two companies of Marines plus the SEALs, was successful in capturing the harbor area. However, snipers from many areas continued to fire at the Marines. The snipers would fire a burst from one window and run away. By the time a larger gun could be turned on them, the shooters were long gone. This did not stop the Marines from coordinating fires from the Super Cobras that were still providing cover.

Next came the LCUs (Landing Craft Utility). They landed with unit integrity, as each carried up to 450 Marines and three M-60 tanks. To protect their arrival, massive amounts of obscurants were fired directly onto the buildings surrounding the port. While the militia did continue firing toward the troops, they could not effectively target the troops or war machines.

The use of obscurants also prevented the Marine use of optical sensors in the immediate area. However, they had new acoustic sensors that could track the flight paths of bullets. In a full-fledged firefight the input was simply too high to discern a single target. When there was a lull, they could locate snipers and provide sufficient data so that machine guns could return the fire in less than a second. In the inventory were small UAVs that were under the control of local commanders. These would be flown through the obscurants and provide pictures of the other side. While an occasional vehicle was spotted, most of the people were hidden inside buildings.

Also available now were Light Attack Vehicles (LAVs), some equipped with TOW missiles and their 155 howitzers. For now the artillery would be used in the less-favored direct-fire mode. It would be effective for countersniper fire from the high-rises. Members of the Air Naval Gun Liaison Company (ANGLICO) support would be used for counterbattery fires. They would live up to their motto, "Lightning from the sky. Thunder from the sea." As expected, within thirty minutes of the assault artillery fire from the mountains above the coastal plain began. The next step would be consolidating the brigade beachhead and preparing for the next phase of the operation.

The invasion near the airport went far more smoothly. Since the first mile was already in friendly hands, the landing craft did not have to fight their way ashore. In short order the Marine brigade and Ranger battalion had moved to the northern and eastern perimeter. Lead elements of TF Sabre were already en route. However, they would be landing under sniper fire and would still be exposed to periodic artillery fire, a situation not favored by the Air Force Transportation Command. In preparation for the FCS arrival, one company of Rangers was sent to the southeast to watch for militia members with HN-5s. Mortars were set up for fire suppression if people were spotted on rooftops. These elements also deployed their small UAVs to circle the area just before the first C-130s began arriving.

For two days the planes landed and took off. Several were hit by small arms fire, but all managed to evade the occasional HN-5, usually fired as they arrived. One unlucky C-130 was hit directly by an incoming artillery round. All aboard were lost. To clear the critical runway a tank pushed the remaining pieces into a nearby drainage ditch.

Abroad there were a number of events that related to the American intervention. Saudi Arabia, which was both actively supporting the various militia groups and selling oil to anyone with money, announced they would dramatically cut back production. They feared an Israeli strike and knew that some of the petroleum products were assisting us in the war effort. As a result, in the United States prices rose at the pump to above five dollars a gallon. That in turn would add to the cost of goods and increase the prices in stores and decrease sales.

Concurrently, there were seven suicide bombings in U.S. shopping malls in just two days. Taking events to a new level, one such bombing in Detroit was performed by a woman pushing a baby carriage. Witnesses who had seen her before the explosion confirmed there had been a live child with her. The bulk of the carriage provided extra room for explosives, and the blast in the center of a multistory mall nearly collapsed the building.

These attacks were designed to frighten average people and eventually weaken the economy. The paradox was that terrorist leaders were orchestrating these apparently random events, yet they did not direct any specific attack. Rather, these executive agents with no governmental affiliations merely described conditions that might be useful to their cause. Individual cells, some of which had been sleeping for two decades, then selected the time, place, and type of attack they wanted to execute.

King Abdullah II of Jordan, one of the few remaining moderate voices, denounced these activities and declared his country neutral. Within two weeks he would be deposed. Another theocracy in the making, the vacancy brought about a brief power struggle, but in the end the mullahs appointed their own leader. While Jordan was not officially in a state of war with Israel, so many forces were posted along the western border that they could not be ignored. Another concern to everyone was the loyalty of these troops and who, if anyone, was commanding them.

Except for Syria, no government had troops in direct conflict with Israel or the United States. That conundrum made diplomatic efforts nearly impossible. The presence of Syria in Lebanon was not a new issue, since they had maintained troops there for decades. Of course the rhetoric from Damascus inferred that they had interceded to prevent the slaughter of innocent noncombatants. Memories of the September 16, 1982, massacre at Sabra and Satilla would never be erased, and they still held the Israelis responsible as they had allowed the killing to take place. In fact, there had been other such incidents as well, like the one at Qana in April 1996.

When Col. John Warden first articulated his five rings of warfare theory, he placed great emphasis on attacking the enemy leadership or causing others to pressure them.[1] His brilliant air campaign for Desert Shield demonstrated the efficacy of that premise.[2] In this conflict, however, it seemed impossible to firmly identify the real leaders. In fact, much of the enemy leadership came from concepts, not people, and the adversary was acting more like a superorganism than a geopolitical entity. When local leaders could be identified, they were attacked. The confounding factor was the number of small units acting semi-independently and generally without a centralized command and control system. This was not like fighting a traditional army.

Once in place, the Army and Marine units began to search the terrain in their immediate areas. From the start casualties were mounting. More than 300 troops were lost during the entry phase. Snipers or random artillery hit most of them. It was fortunate that their improved body armor stopped most of the bullets and fragmentation. Automated tourniquets activated by biosensors improved the survivability for those wounded on extremities. The fatality rate was held down thanks to advanced medical attention, including the availability of synthetic blood. Prompt evacuation to medical stations and out to the waiting hospital ship where surgeons could operate within the golden hour helped tremendously.

With troops prepared for urban battle, the attacks began on D + 3. In the north two Marine battalions crossed Shari Barie Road under cover of darkness and immediately ran into trouble. As they approached the first buildings, they began taking fire from the front and flanks. Just a four-block front was the extent they could cover in their attack. Before the assault they had spent hours visually reconning each building and noting the locations of snipers. When the shooting started, the Marines found the snipers were constantly moving. Experienced in their trade, the shooters stayed well back in the rooms from which they fired. Flashes were rarely spotted. It was also obvious that the militia had some form of night-vision equipment. While not as good as the third-generation U.S. goggles, it was sufficient to spot movement and fire. Even as they were crossing in their LAVs, there was no alternative for the troops but to be exposed in order to assault the buildings. They could not be protected from snipers shooting from all angles. Crossing open areas is known to be the most dangerous situation when fighting in cities.

Breaching buildings was also a serious venture. Going in the front doors could be even more dangerous. So special munitions were available to blow holes in sidewalls. It was evident that the heavy direct fire weapons would not be of much use for opening buildings or countering snipers. Soon all of the American forces would learn what the Russians had in Grozny: turning the building to rubble only provides the enemy more hiding places. The growing concern of commanders was that the casualty rates might be commensurate. Urban fighting is known for very high casualties on all sides.

In two days TF Sabre had advanced about one hundred meters in a limited sector. NetFires, the long-range systems were useless except for counterbattery fire and interdicting supplies moving in the Bekáa Valley beyond the mountains. The robots provided the ability to put sensors in sensitive areas, but they, too, encountered competition. Very small robots were good when they could hide. As more and more rubble accumulated, they experienced increased difficulty in maneuvering about. When militia troops got close to them, they simply threw a blanket on top of them. Most useful in sensing were the micro-UAVs that could hover and peer into windows on any given floor.

The buildings to the north were even taller than those initially encountered. It was very apparent that additional force would be required if the mission was to have a chance for success. Central Command received a request for another

Ranger battalion to be sent in immediately just to hold on to the current gains. If they were to pursue the mission as originally outlined, it would require multiple infantry divisions.

The National Command Authority was now faced with a serious inherited problem. The requested divisions did not exist on active duty. With a press to get new technology into the military and transform the way they fought, the number of people was cut to a dangerous and unacceptable level. To save money an attempt had been made to decrease the active force below eight divisions in the Army and two in the Marines. This transferred the majority of the combat arms into reserve units. The old argument that kept the tooth-to-tail ration in favor of combat power up front had lost. At the same time, a paradox had emerged in warfare. The speed of conflict increased, as did the complexity of weapons systems. Conversely, the troops had less prior training than their active-duty counterparts, thus requiring greater time to prepare once called to fight. Further, Military Operations in Urban Terrain (MOUT) was even more dangerous than traditional warfare on open terrain and thus required extensive and current training. Despite paper reports indicating a high state of combat readiness, it was simply a bridge too far. The willingness of commanders to please the Pentagon with politically correct estimates would cost many lives if their units were thrown into combat prematurely. This was not peacekeeping; it was war. They had bet on the outcome—and lost.

Also to be considered were the economic implications of reserve call-ups. As experienced repeatedly, when units that are geographically located are activated the towns from which the people come can be greatly impacted. Professionals' taking substantial salary cuts and families' loss of one or both parents take money from the local economy. An answer to the complex problem lay in revamping the military personnel system. Skilled retirees abounded, most willing to be re-called to fight if an urgent need existed. There was no need to have members of the military serve twenty years consecutively. Over the decades a small number of reservists had proven that intermittent periods on extended active duty could be effective and they could keep up with their counterparts. A strategic response would be to move retirees into homeland security positions and push active-duty personnel into fielded forces abroad. That would come later; right now the White House had to decide about reinforcing units in Beirut.

The conflict had to be refocused in order to break the impasse and yet retain our position of relative importance in the world. That would take time, and the 80th and 98th Infantry Divisions were ordered to active duty and to prepare for deployment to the Middle East. But it would take at least three months before they were ready for action.

In the New York Times *Science Section an article ran that was to change the game. It described not a new weapons system but rather the successful engineering of a device that actually ran on zero point energy (ZPE). Decades before, a brilliant and innovative theoretical physicist, Hal Puthoff, Ph.D., had published an obscure article in* Physical Review D *proposing that ZPE existed and could be extracted.[3] While empty space or a vacuum was thought to be a void, quantum theory suggested otherwise. Empty space is considered to be a vast domain of random, fluctuating quantum processes that provide a sea of energy.[4] Further, Puthoff asserted that once engineered, the amounts of available energy would be astronomical—and anyone with a generating device could tap into the nearly infinite source. If true, such an engineering breakthrough could alter the geopolitical balance of the world. No longer would people need to rely predominantly on petroleum products for energy.*

For years Puthoff and his colleagues suffered derision at the hands of most conservative scientists. But with modest private funding, the proponents of zero point energy had made significant gains. Now they were publicly announcing the successful transition from theoretical studies to actual engineering. From a practical standpoint it would take about eight years to fully integrate ZPE into standard use. The process would be accelerated because of the current conflict and the pressures it brought on the world oil market.

Of course none of the militia fighters read the article or would even give a damn if they could. However, a handful of scientists, economists, and political leaders in Riyadh, Tehran, Damascus, and the smaller Gulf countries understood the implications. The world as they knew it had changed. The requirements for their petroleum products would diminish. Slowly at first but at an accelerating pace the price of oil would drop. Within a decade most of the Middle East could become of no consequence to the technically developed countries. As much as the fundamentalist religious leaders despised Western ideals, there were other powers that influenced policy—money being one of them. Wars cost lots of it, and the source was about to dry up.

Beirut remained a problem as slowly the war dragged on. Units employed the latest in technology. Whenever possible a platoon would land on the roof of an objective building and fight their way down. These attacks were tricky as helicopters were vulnerable to the HN-5s. However, the new millimeter wave Active Denial System (ADS) brought some relief. Mounting the ADS on tall platforms, the base security forces could sweep the surrounding area with the pain-inducing beam. This prevented the militia from using rooftops as air defense platforms. A version of the ADS was mounted on a CH-47 helicopter and used to reach into sectors outside the line of sight of the base. This innovation dramatically reduced the interference with heliborne operations.

Using see-through-wall sensors the troops could determine if people were on the floor below them or in a parallel room. Unfortunately, the sensors could not distinguish between militia and noncombatants. Frequently, optical sensors were used. Sometimes the troops simply drilled through walls or floors. Other times the micro-UAVs would be flown to the windows. There were also minirobots that would crawl into position and often be sent into a room before the assigned squad rushed in. They performed better going downstairs than climbing to the next floor. However, no matter what tactics were used, booby traps within buildings continued to plague the operations. Although very sensitive chemical sniffers were available, the entire environment had become so contaminated with nitrogen products that the sniffers constantly registered explosives.

Improved squad communications systems that provided the exact location of each person were a godsend. One of the biggest fears of the troops was engaging in a shootout with another squad from the same platoon. Considering the rising casualty rate, the soldiers and Marines had become very jumpy. It was not unheard of to have friendly elements engage each other in firefights through walls. At all levels command and control was extremely difficult.

Another problem was ensuring a building once cleared remained safe. There were not enough troops to leave people behind to monitor each floor of every building. Whenever possible, the area was sealed. Fast-curing rigid foams could effectively seal a door or air vent in a matter of minutes. Still, frequently militia members would attempt to sneak in behind the U.S. forces and ambush them in places they had previously secured. Whenever possible, small thermal and motion detecting sensors were left hidden in key areas of those buildings. Centrally monitored, they provided information about disturbances in secure areas. When alarms went off, troops had to be dispatched to check out the situation.

Movement was slow, very slow. At first it often took a company one week to clear a city block. Casualties ran from 20 to 50 percent in some companies. The average range in firefights was less than 100 meters. Sometimes it was only a few feet, and hand-to-hand combat was not unknown. Through terrible lessons they came to understand the admonition of Col. Gary Anderson, USMC (Ret.), who forecast the future adversary. He reported that the kill ratio for Hezbollah against Israel was 10:1 in Hezbollah's favor. Further, in seven years no Hezbollah members had ever been captured alive.[5] In Beirut, if trapped, they would stand and die in place and consider it an honor to do so.

The survivability of American forces improved over that which foreign armies had experienced in urban combat. In addition to body armor, new robotic vehicles played a major role in getting the wounded to safety. These remote-controlled devices could be steered to the location of the wounded person. When the soldier was in the open, the robots could drive over top of him and scoop him into the vehicle. At other times they drove into buildings so that people would not be exposed while loading the wounded.

Similarly, these robotic vehicles became indispensable for sustaining the force. The firefights in the buildings swallowed up immense amounts of ammunition. Resupply was critical and provided a distinctive edge for the militia. In addition to ammunition, food and water were often needed. The energy consumption rate for men in intense combat was extremely high. Water was a critical resource. This robotic resupply saved several squads that became cut off from their parent units and would have run out of supplies. In open combat missions, troops receive support from helicopters or airdrops. Inside the city they are rarely available, especially in areas with taller buildings.

In the United States suicide bombings continued, albeit at a slightly reduced rate. The fact that they happened at all changed American buying habits. More things were bought and sold over the Internet as a means to avoid congregating in public places. In response, malls, sporting events, theaters, and even churches increased security. The ratio of private security to government law enforcement rose to 7:1. Although the quality of service varied based on funding and criti-cality of the assignment, major security firms improved their professionalism. Local law enforcement understood they could not provide all of the protection desired. This led to new, previously unthinkable relationships between private security and government agencies. These entailed far more than contractual agree-ments and included cooperative efforts in counterterrorism.

Growing casualty rates with small advances frustrated the general population. Desert Storm, Enduring Freedom, and Iraqi Freedom had established models for warfare. They were quick and distant and produced quantifiable results. None of this was happening in Beirut. Naval gunfire and bombs were of little use. Occasionally a convoy would be detected in the Bekáa Valley and attacked. If bombers were in the area they did the job. Sometimes the Texas would launch one of its cruise missiles. But in the end, they had little effect on the war, as most supplies came in on single vehicles and rarely were supply dumps located.

It was decided by the National Command Authority that a twofold strategy would be executed. First, militarily, there must be an acknowledged winning situation in Beirut. It would be too great a loss to appear to be driven out as we were after the bombings of the U.S. Embassy and the Marine barracks in 1982. Therefore, rules would be changed and the outcome assured.

Second, the United States would enthusiastically pursue alternative energy sources on a scale previously unparalleled. No lip service this time. Energy drove world politics and it was time for America to reassert supremacy. The development of devices employing ZPE would get the highest priority, leading to independence from foreign oil. In addition, the United States would make it clear to the technologically developed countries that an alliance with us would mean inclusion in the energy paradigm transformation. That carrot should be too great for any country to resist.

The military operation was to be split into two separate efforts. One belonged to Special Operations Command. There would be a series of clandestine operations deep in hostile territory. Instead of destruction, leaders of the opposition would be targeted for abduction. While the amorphous adversary did not have a designated command center, there were several high-visibility mullahs functioning in public and issuing the call to arms. They did not assign targets but rather announced what was appropriate and allowed the independent groups to choose the specific sites and carry out the attacks. Holding a number of these mullahs would both provide a bargaining tool and demonstrate the ability to counterstrike in unexpected ways.

The second aspect would be to alter the fight in Beirut. When the chemical weapons convention was signed in 1993, the signers did not foresee hamstringing the military to the degree it did.[6] For altruistic purposes there was an international hue and cry to ban chemical weapons, including riot control agents (RCAs). Though such weapons were still permitted for domestic and non-military situations, the treaty prohibited their use in combat. To cover their butts, lawyers often made decisions that had strategic impact based on such outmoded constraints.

Lost in the argument was the experience of the 1st Battalion, 5th Marine Regiment, and their desperate battle to regain control of the Citadel in the ancient city of Hue, South Vietnam, after the Tet Offensive of 1968.[7] The battalion was assigned the mission to capture an area ten blocks deep. Due to its cultural value and religious significance, restrictive ROE were in place. Therefore, very limited air strikes or heavy artillery would be available for supporting fires.

After three weeks of intense fighting the Marines had advanced only three blocks and suffered horrendous casualties in the process. Fighting took place with enemies just a few feet apart. The Ontos, a vehicle with six 106 recoilless rifles, fired round after round at point-blank range with little effect. Nothing seemed to work. Then the Marines were provided with a new weapon, the M8 gas launcher, which fired forty 40mm CS rounds in a single burst. Only three were *available for the assault. However, once employed the CS gas (a powerful lach-rymal agent) cleared the entire area. Following the blast of gas, the battalion was able to cover the remaining seven blocks without opposition. In short, this non-lethal agent had broken the resistance of the hardened North Vietnamese Army fighters and undoubtedly saved the lives of many Marines.*

Both riot control agents and calmatives had advanced a great deal. Despite the incessant complaints from peace groups, there was sufficient legal justification to continue the research. The National Command Authority determined that saving lives was worth the risk and announced that the United States was reconsidering their position on chemical agents as they pertained to life-conserving applications. In short, we were going to use non-lethal gas. While the agents were not perfect, they constituted a huge leap over the war of attrition that had bogged us down. Not mentioned but authorized in the same directive was the use of anesthetics against designated targets.

With the public announcement the perception campaign rolled into high gear in both camps. Since the initial incursion, the militia had been waging a very effective operation aimed at winning the minds of neutral countries and dissi-dents within the United States. There had been many articles filed with every newspaper on the planet. The Internet proved to be a useful tool as Web sites supporting the militia's position sprang up by the dozens. After every attack there were pictures of unfortunate children who had been wounded or killed in cross fire. Many of the sites masqueraded as legitimate concerned citizens groups when they were actually under the absolute direction of the militia.

Al-Jazeera again became an international media source. Virtually unheard of in the West prior to Enduring Freedom, this Qatar-based television network became a global source of pro-militia stories. Early on they had invoked the wrath of the House of Saud; thus when the fall approached they were instru-mental in tipping the scales. The power of television was well understood by leadership on both sides. Plagued by young correspondents out to make a name for themselves, the nearly unrestrained Western reporters daily filed a huge bar-rage of stories, hoping that one of them would be their ticket to stardom. The frantic pace made fact checking an ancient art. Many times these reporters were duped into printing erroneous reports. The reality was that both sides manipu-lated the news for their own purposes. With casualties rising, Americans, with their traditional short attention span, were quickly losing interest in foreign military adventures. With a profound lack of critical thinking skills, that pop-ulation listened to sound bites for explanations for complex issues.

In mid-2002 the U.S. Special Operations Command achieved a degree of autonomy for assigning and conducting missions in theaters around the world.

*Now it was time to initiate one of the most dangerous and multifarious oper-
ations ever conceived. It entailed the coordinated abductions of key figures in
several countries. Each person had high visibility in his own right. That meant
he had either personal protection or followers who would be in close proximity.*

*Snatching these figures would be far easier than successfully extricating them
and moving them to U.S. custody. The timing would be critical. If word spread
of the disappearance of leaders in one city, others would increase security or
move to inaccessible locations. The people now known as packages were located
in Damascus, Riyadh, Amman, and Alexandria.*

*Access is the top priority for SOF personnel. The ability to be where they need
to be, when they need to be there, is what they called access. While they were
trained for various military insertion techniques such as HALO parachuting,
helicopter raids, or even swimming, often using the front door was the simplest.
In this case, arriving on commercial airliners was the insertion that would raise
the least concern. Over a period of a week each soldier arrived alone. Most were
indistinguishable from the indigenous population. A few would emulate European travelers on business in the area. All could speak the language of the country
they represented as if they were natives. In other areas of the world they might
have arrived with a female and acted as a married couple, but given the role
of women in these societies it was not practical for this operation.*

*In every case, this was not their first trip to each destination. Part of the
SOF background was to travel to areas that could become operational areas and
become familiar with the terrain and customs peculiar to those specific locations.
The teams brought limited equipment, all disguised as innocuous baggage that
could stand up to a thorough search. Most sensitive were the secure communications that could be assembled from components of laptop computers every
businessman carried.*

*For the following days each agent would travel about the town holding meet-
ings, sightseeing, or shopping. In reality they were reconning the area and picking
up information about the location and accessibility of their package. Rarely were
more than two team members together for any length of time. They took turns
over-watching other team members to determine if any of them were under
surveillance. Had a tail been spotted, a signal would be given and that member
would exit the country.*

*The Joint Special Operations Command Headquarters was responsible for
gathering intelligence reports from national collection sources and coordinating
the operation. The intercepts from the National Security Agency were compared
with CIA human intelligence reports and the information sent back from the
SOF teams with their boots on the ground. The use of sophisticated, secure*

operations planning models allowed for analysis of many extraneous bits of in-formation searching for the best combination of situations. Success meant that at least two of the teams had to be able to conduct simultaneous snatches and exfiltrate with the packages. Three concurrent abductions would be excellent, and the probabilities against four successful raids were astronomical.

Damascus proved to be the most problematic. The targets moved constantly and there was no ability to predict where they would be at any given time. They were cautious with their few phone calls, and the National Security Agency could not provide any useful information. After ten days, the team was signaled to abort the mission and leave the area. In the next two days each member boarded a commercial airliner headed for a different destination. Each team member would have at least two intervening stops before returning to the United States.

Luck held for the remaining teams and soon the computer model detected the desired coincidence. Following the evening prayers on Friday each target was headed to a small social gathering. The locations were guarded but not inacces-sible. No more than ten people would be at any of the locations, a number that easily could be overwhelmed by the four-man SOF team.

On Thursday night support teams were inserted by parachute at the assigned exfiltration points in Saudi Arabia and Jordan. Their mission was to ensure protection of the sites and provide extra firepower in case the operatives were discovered accidentally during or after the abduction. Alexandria would be a sea operation executed with an Advanced SEAL Delivery Vehicle that would return to the USS Texas, *which was snuggled in the Khalij al-Arab not far off the coast.*

Each snatch was different and yet strikingly similar to the others. Three SOF soldiers walked within a few feet of the objective area. The fourth brought a vehicle within a block of the site, stopped, and looked under the hood at the engine. Since maintenance was not a high priority and vehicles tended to be rather old in this part of the world, engine problems were common and provided cover for stationing the car in this location. With precision timing the agents approached the site from different directions. A guard was distracted momen-tarily by one agent while another fired an anesthetic dart into his leg. In ten seconds, and without sounding an alarm, the guards were put down and moved out of sight.

Donning miniature breathing devices, the agents then opened the door and threw in grenades containing calmative chemicals that were dispersed throughout the room. With silenced pistols, the agents entered the room. Two people, not fully incapacitated, attempted to resist. They were dispatched, as there was no time for fooling around with extra prisoners. Facial identification checks verified the principals, who were bagged and bounded. The agents departed knowing that this group would be unconscious for several hours. When they regained their senses it would take another two hours for them to collect their thoughts and convey what had transpired.

In under ninety seconds the vehicles arrived and the groups were headed for the designated recovery points. Erratic driving was the norm in these cities, so darting through traffic would not attract attention. The biggest concern was being involved in accidents, as most drivers there were not licensed and their skills quite limited.

Turning into the desert miles outside Amman and Riyadh, the teams began signaling with infrared devices. Tilt-rotor Ospreys were flying at full speed just off the sand and would arrive at the landing zones shortly after the teams arrived. The landing zones were marked with infrared markers as a low-probability-of-intercept beacon directed the MV-22s to the sites.

Waiting at a nearly deserted pier in Alexandria, a SEAL team had boarded several small boats and set up machine guns to cover the raiders. They, too, spotted the IR signals and prepared to move out to sea. Three fishermen stumbled into the operation by accident. They never saw the SEALs who slipped up behind them with darts.

At JSOC the message came in. All of the packages had been delivered safely. The Special Operations forces had again played a strategic role in resolving this conflict. The names of individuals would not be known, but their deeds would.

As Sunday evening fell in Beirut there was little to distinguish it from any other. Sounds of sniper fire punctuated the darkness. Small cooking fires dotted the area, and occasionally the figure of a person was spotted darting quickly across an unlit street. The militia had not been as active in the past few weeks. On the Islamic radio station there were fewer cries ranting against the infidels. The satellites and UAVs had indicated that traffic to and from the Bekáa Valley had slowed.

The Pentagon took credit as a rationale for the expenditure of precision-guided munitions against the infiltration routes. As had happened early in the war, the inventory of PGMs was running low. While popular with the military and the press, these were expensive munitions and required lead time to manufacture. Too frequently the military was slow to order, causing fluctuations in the work-forces at Raytheon and Lockheed. The budget analysts never did understand the full impact their inconsistencies brought. All they worried about was whether their curves for expenditures looked right on the briefing charts.

At three in the morning the attack began. The U.S. forces moved out from their bases toward several key facilities in the city. As before, they heavily em-

ployed the obscurants to cover their movement. Unlike in previous operations, combined with the smokes were concentrations of CS gas. The masked American forces were protected from the effects and their superior night-vision equipment would allow them to have better visibility than the militia members.

Shortly after the non-lethal agents were dispersed, coughing could be heard coming from the nearby buildings. Soon people began pouring into the streets wheezing and holding cloths over their mouths and noses. This was ineffective in stopping the effects of the CS. Many children were crying and mothers attempted to lead them to relative safety. Some barricaded themselves in cellars, but that would be a temporary measure.

In fact, the CS agents had not been widely employed. Rather, they were concentrated in key areas and along the streets the troops would travel. Raiding teams were assigned to hunt down militia members and other soldiers, and Marines led women and children into areas that were free of the gas. Any building that seemed sealed received special attention. Robots would prowl the halls sensing life-forms. When one was located a fire team would check it out. If in doubt, the team would puncture the door and insert a fast-acting calmative agent. Thirty seconds later, they would blow the door and secure the room. The temporarily incapacitated occupants would be sorted out. Fighters would be cuffed and taken to the street. There they would be placed on the trucks that were sent through behind the fighting forces.

Occasionally as the CS gas began to permeate the redoubts, militia members would fire wildly in the direction they thought contained Americans. Unfortunately for them, U.S. troops could see quite well and used aimed fire to neutralize the threat.

By dawn the American forces had cleared and occupied more buildings than had been done since the invasion began. More than 600 militia members were prisoners, and several hundred more lay dead. Many of them had killed themselves to avoid dishonor. Some used grenades and tried to kill the American troops at the same time.

Before noon the utilities and transportation hubs were all in American hands. The prisoners were moved back to the airport, and aid stations were assisting those complaining about residual effects from the CS gas. Although two elderly people with lung infections had died, the collateral casualties were lower than in any previous operation.

After consultations between Washington and Tel Aviv, back-channel signals were sent to all governments in the Middle East. If the terrorists' bombings would be halted, Israel and the United States would pull out of Beirut in short order. The rapid success of the U.S. military operation in the city had stunned those supporting the various militia factions. In addition, they had word of the important mullahs now being held on ships of the Sixth Fleet. They and the prisoners from Beirut would be traded for American and Israeli soldiers held captive. Given the unexpected advances in energy generation, economically astute leaders conceded that a major geopolitical shift had happened. Ideology aside, it was in their best interests to reconsolidate and prepare for the emerging world. For the Americans it was time to declare victory and gracefully leave. As with Vietnam, for decades to come it would be hard to justify the loss of so many fine men and women.

ARGUABLY THIS IS THE most important chapter in this book. It elucidates complex issues that are inextricably intertwined. While many military leaders and observers comment about the importance of urban combat, they have not been successful in causing the dramatic shift in thinking necessary to move the Army to make MOUT the major concern. Certainly the U.S. Marine Corps has been more forward-looking in this matter. The dominance of urban population in the littorals made the mission obvious to them. For years the Army wanted to bypass cities to fight on the open terrain. In fairness, there have been voices in the wilderness crying out the warning. Strapped for resources, planners have devised tactics by which they can attempt to dominate cities by holding critical objectives. This approach may work for a quick raid, but it belies the lessons learned by every other army that has engaged in long-duration urban combat against a determined adversary that is willing to accept casualties.

Most of the military's new weapons are designed to increase standoff distance between opposing forces. In urban fighting, ranges often are limited to tens of meters, not kilometers. We need more emphasis on weapons that are effective when the fighting is up close and personal.

Among the critical issues raised here is the lack of adequate forces available for major conflicts that require ground troops. The current doctrine requires activation of reserve components on a regular basis just to meet relatively minor conflicts. While these service members are prepared to serve when exigent circumstances are prevalent, repeated activation will foster problems of retention and reduced support from employers. Most of these people have day jobs, and active duty detracts from the local economy. Urban combat requires extensive training and will not be available to all recruits, yielding increased casualties. Technology alone will not solve the problems associated with urban combat. It requires boots on the ground, and current planning is a recipe for disaster.

Under current treaties even non-lethal chemical weapons could not be employed legally in this scenario. The paradox is that their use could save lives of combatants on both sides and non-combatants trapped in the area. As U.S. casualties mount, Americans will demand all measures necessary be taken to reduce them. To prevent a future crisis, chemical warfare treaties should be revisited immediately.

Energy is key to global politics. The chapter titled *Six Sigma Solutions* will explore some of the technologies that can break our dependence on petrochemicals. Developing new energy sources is indeed an asymmetric response to current and future conflicts.

PART III

PLAN A: WIN THE WAR ON TERROR

THE WAR ON TERROR has become the rallying cry for nations that profess Western values. Enduring Freedom in Afghanistan demonstrated the ability of the United States to cobble together an unlikely coalition of perennially feuding tribes supported by overwhelming airpower and geographically defeat a specified adversary. Militarily the Special Operations forces conducted magnificent guerrilla operations and effectively directed precision-guided bombs to their targets. In short order they accomplished what the Soviet Army could not in a decade of effort and at horrendous cost in lives. The attacks also required phenomenal diplomatic efforts for basing and overflight rights. The fundamentals of Enduring Freedom set the tone for new conflicts—at least in the near future.

Concurrently these measures prescribed eventual failure. The U.S. military and their allies can and will fight effectively. However, these efforts will be in vain because of the lack of understanding of the social construct of the era we are entering. Again, Afghanistan serves as an example. The fundamental question is why Afghanistan? For millennia that territory has been one of the most inhospitable areas of the globe. It is populated by warring tribes, who have no interest in a centralized society with common values and purpose. For reasons inexplicable, outside elements have repeatedly attempted to forge these fiercely independent people into something they are not—a nation-state that acts in a unified manner.

In the United States great enthusiasm was shown as one Afghan faction after another seemed to come over to our cause and rose up against the Taliban. The reality was they did not join our cause at all. They followed the age-old values of the area—taking the money and joining the winning side. This was not a moral victory, only temporary pragmatic alliances of convenience. Even before the Taliban had been officially defeated, old animosities resumed. Special Forces troops on the ground were on constant alert to ensure they did not call in air strikes on another team supporting a different tribal faction. The problem was so common it became known as Green-on-Green. That meant the U.S.-led coalition was providing support to both sides of a given skirmish.

For the near future we will continue to attempt to isolate potential adversaries by defining them as terrorists. We will then provide technical and military support to local groups that seemingly support our wishes. Bribery will be a normal part of both the intelligence and operational processes. In addition, enormous amounts of funding will go into providing indirect support so that the coalitions will not unravel too rapidly. These costs will likely remain hidden from view, so that the true total costs for the War on Terror can never be determined. Those costs would simply exceed the tolerance level of the taxpayers, and they would rebel against these unfocused efforts that proliferate around the world.

The fact of the matter is that we do not have much option other than to execute Plan A with the hope that it can forestall the inevitable. Our Judeo-Christian-derived values would not permit us to initiate the level of violence necessary to eliminate terrorism and again become a dominating force throughout the world. Our response to terrorism will be pain-mediated. Paradoxically, if the terrorists succeed in attacking U.S. interests and killing substantial numbers of our citizens, it will be the painful catalyst required to ensure their destruction.

The United States is the world's economic leader and certainly has the most technologically advanced military capability. However, our military lead was severely endangered during the Clinton administration. The damage will take many years to repair and the current path will be insufficient to overcome the loss of momentum. The Department of Defense technology base was moribund too long and a bigger boost is needed. However, in the end it will be people, not machines, who determine the outcome of war.

Future War described the emerging threat. It was accurate then and continues to be so today.[1] Many in the military have spent too much effort preparing for the war they want to fight and too little time preparing for the wars they fight. Of course, some others in the military have long recognized this conundrum. Unfortunately, they have not prevailed.

Secretary of Defense Donald Rumsfeld rightly notes that our ability to adapt will be critical. Our challenge, he stated, would be to defend ourselves "against the unknown, the uncertain, the unseen, and the unexpected."[2] We have, however, set off to prepare in a manner too shortsighted, even with the increases in spending that have been allocated.

In this section we will explore the inevitable aspects of Plan A. It will be thrilling as advanced technology provides capabilities unimagined only a short time ago.

CHAPTER TEN

THE ROOT OF ALL EVIL

"FOLLOW THE MONEY." THE words of Watergate confidential informant known only as Deep Throat are as wise today as they were thirty years ago. The difference is that the complexity of money-handling schemes has increased dramatically. Electronic transfers happen with a flick of a computer key. Money moves internationally through multiple accounts with blinding speed. Given the near trillion dollars generated annually by organized crime on a global basis, there is controversy about when illegitimate money becomes legitimate. Add to that the money used to support terrorism and the money sleuths have a daunting task.

The only way in which the War on Terror can be won is by eviscerating the funding channels. Unfortunately, we will not take the actions necessary, as there are too many sacred cows and ethical issues involved. As will be seen later, the failure to take difficult measures now will result in a far larger crisis in the not too far distant future.

While terrorists may be physically attacked when they are located, the real key is to eliminate the support base that provides them with weapons and other means to attack us. Everyone has heard about the financial capability of Osama bin Laden and how he applied his wealth to developing the al Qaeda network. It is also known that his personal fortune could not have created an organization of that magnitude. Rather, he had help from the bin Laden family members who continued to support him financially even after the 9/11 attacks.[1]

In order to stop the money flow certain steps need to be taken. While I do not believe we currently have the will to execute them, it is essential that they be examined. As more pain is inflicted upon us it may become necessary to reconsider these proposals. The remaining question is whether we will take action before it is too late.

STEP ONE

In addition to private funding, many of the terrorists' activities are supported by illegal activities. Foremost among these is the illegal drug trade. Illegal drugs provide the largest source of terrorist funds, an estimated $300 to $400 billion per year. The drug trade has been estimated to cause up to 80 percent of the crime against property and 50 percent of the violent crimes against people in the United States.

Despite extensive efforts, including spending billions of taxpayer dollars attempting to cut supplies, the street price for cocaine and heroine remains about constant. From that we know that even with some highly publicized successes in interdiction, the supplies are coming through on a regular basis. At times enforcement has caused traffickers to shift their patterns. Processing bases moved from the Amazon jungle to the Pacific coast. Routes of shipment and ports of entry change periodically. More innovative methods for smuggling are developed. And still the drugs keep flowing.

To state "we are winning the War on Drugs" is the equivalent of Saddam Hussein stating that "the mother of all battles has yet to start." The war, if there ever was one, was lost before it began. The demand in the United States was greater than the desire to stop it. In the process we have seen whole countries corrupted, regions destabilized, and even threats made to our own national defense through loss of trust in selected units and reallocation of resources. The money is just too big to be stopped with conventional measures.

Using Gen. Barry McCaffrey's 1999 figures for the price of cocaine, the problem becomes readily apparent. In his congressional testimony he estimated the production cost per gram to be $3 and street price between $150 and $200. That means that a kilo of cocaine cost about $3,000 to make and will sell for up to $200,000. That is a profit margin of 98.5 percent, and only because the product is illegal.[2] There is no other product in the world that has a higher return on investment. Futility in the supply-side approach is noted later in McCaffrey's testimony. With pride he indicated that production of cocaine in Colombia and Peru had dropped by 3,000 metric tons. The street price for that amount would be about $600 million. Despite the reported drop in production, suppliers could keep the price constant. It is worth considering what size an industry must be to absorb a cut of that magnitude without blinking. Balanced against the global income from illegal drugs, estimated at about $400 billion, that decline in production is statistically insignificant.

McCaffrey's congratulatory remarks should also be viewed from a cost-benefits standpoint. The 3,000 metric ton reduction was over a four-year period. In 1999 the Office of National Drug Control Policy (ONDCP) budget was $17.7B.[3] Allowing for increases each year, ONDCP expended over $60 billion in the War on Drugs over the four years and accomplished a $600 million decrease in product. Even at inflated street prices, that is a 1 percent return on investment. By 2001 the budget had risen to $19.2 billion with constant market prices. When the ONDCP budget, since it was formed in 1988, is combined with the cost of increased incarceration, American taxpayers have contributed over one half trillion dollars. Considering there has been no discernible decrease in availability or street price, the zero return on investment should be unacceptable. Obviously there is something wrong with the math.

It is easy to blame Colombia, often the butt of comedians' jokes, for supplying cocaine to the United States. The reality is that if we had suffered the same percentage of casualties as have the Colombian military, government officials, newspaper personnel, judges, and others who have opposed the narcoterrorists, there would be a hue and cry for an armed response and willingness to forgo protection of civil rights. Imagine what we would do if the U.S. Supreme Court were physically taken over by traffickers. Few Americans remember that it happened in Colombia in 1986 when the Medellin Cartel took over the Palace of Justice, killing ninety people, including eleven Supreme Court justices. Later, in 1989, the leader, and reportedly then seventh richest man in the world, Pablo Escobar, was responsible for assassinating three of five presidential candidates.[4] Before he died, Escobar was responsible for thousands of deaths in Colombia, including those of 107 people on a bombed Avianca airliner.

Such violence would be unthinkable in the United States, and if it occurred, we would have responded vigorously. In short, we should be sympathetic to the southern refrain that states, "When Yankees stop shoving it up their noses, we'll stop sending it."

The issue is not one of supporting drug use. In fact, there is evidence that demand reduction and treatment are likely to be more effective than interdiction. As a minimum, two significant goals will be accomplished by legalization. First, and foremost, a major funding source for terrorists will be cut entirely. Second, there should be a steep decrease in secondary crime attributed to theft and robbery to support expensive drug habits and gang fights over distribution rights of their products.

Many books and articles by prestigious authors have made similar arguments. Economist David R. Henderson, Ph.D., addressed the issue of support for terrorism through continuing the War on Drugs.[5] In that article he noted how the Revolutionary Army of Colombia (FARC) financed their terrorist campaign to overthrow the government with drug money. Members of the respected CATO Institute have written extensively about these problems, as have many others.[6] If the terrorists' financial base is to be attacked and eliminated, it cannot be done without removing the financial incentive of illegal drugs.

Contrary to the emotional appeals from ONDCP sources, legalization is a pragmatic approach to shaping the battlefield. During Desert Storm, General Schwarzkopf did not attack straight into the prepared Iraqi defensive positions. Rather, he enveloped them with the famous Operation Left Hook and attacked the deployed armored forces from an unexpected direction. Before that, initiating Desert Storm he saw to it that the battlefield was prepared. Only then did he attack, yielding a decisive victory. So, too, must the War on Terrorism be fought from a battlefield that is prepared and not attack the enemy's strength. Pouring more money into a lost cause will not help. In fact, continuing the War on Drugs will actually diminish the chances for success in the War on Terror.

STEP TWO

The second step that should be taken to cut terrorists' financing will be equally unpopular, but for other reasons. We should let the price of diamonds float to the natural price based on geological availability or industrial capability. In a short time this action would stop the illegal sale of blood diamonds, or conflict diamonds, as they are sometimes known.

While De Beers wants you to believe that "diamonds are forever," they only mean the stones will last for a very long time. What the company does not want you to know is that they artificially control the price and that the value of those precious rocks could plummet instantly to that of glass. The rings and necklaces will still look good, but the man-made gems would probably be even better. Neither the oil nor drug cartels can boast the success rate of the international diamond merchants in maintaining artificial value.[7]

Diamonds have been found on every continent. While Africa is best known for diamond mining, these mines were not located until the 1900s. India was once the chief source of diamonds, and in 60 A.D. diamonds were discovered in Borneo. Actually, Brazil led the trade from the 1700s until yielding to the African riches. Russians have been mining Siberian diamonds and acquired a substantial reserve. In the 1970s diamonds were located in Kimberleys, Australia. In a coincidence, the town had that name before the diamonds were discovered and had no relationship to the famous mines of the same name in South Africa. In the late 1980s intense exploration for diamonds in Canada was undertaken. Geologists located numerous diamond-bearing kimberlite tubes north of Yellowknife in the Northwest Territories. Here mines with millions of carats of diamonds have just begun opening. Their location is so remote that production costs can only be borne if the price remains high.[8] Kimberlites also have been found in Alberta, Saskatchewan, Manitoba, Ontario, and Quebec.[9]

More recently processes for making synthetic diamonds have been developed. Mainly used for industrial purposes, they are harder, heavier, and better heat conductors than their natural counterparts. It is likely that these higher-quality stones may soon be available for jewelry as well. There are indications that they will be indistinguishable from the mined variety, thus making high-quality gems available on demand.

Even so, between the Russian reserves and the Kimberley mines of South Africa there is a sufficient supply of diamonds to flood the market. If any other product were held hostage by a few sources, it would be considered illegal. In the United States there are laws against monopolies and market manipulation. Because of the power accumulated by the diamond companies, they have remained outside the scope of these laws.

The deaths of thousands of Africans can be directly attributed to De Beers price manipulation. When a map showing geological distribution of diamonds is overlaid with a map of turmoil on the continent one can see a high correlation. While many issues underlie the tribal rivalries and ethnic hatreds, it is the lure of diamonds that is a common factor. Former ambassador and undersecretary of state David C. Newsom stated that "stability in Africa will ultimately depend on some effective method to curb such exploitation."[10]

The income from blood diamonds does not approach the amount of wealth provided by illegal drugs, but it is still sufficient to support terrorists and fuel wars. Due to the nature of the trade it is impossible to determine how many illegal diamonds are smuggled. It was estimated the over 4 billion dollars' worth of such diamonds were sold after sanctions were imposed on UNITA as part of a peace agreement between Sierra Leone and Congo.[11]

Direct ties between al Qaeda, Hezbollah, and other terrorist groups and blood diamonds have been addressed by others.[12] The legacy, it was noted, is over one-half-million lives per year and billions of dollars that these wars costs. The problems associated with the diamond trade, both legal and illegal, have been described in articles and books and on television programs. As early as 1982 the *Atlantic Monthly* ran extensive articles describing how diamond prices were being manipulated.[13] The information gathered for a television documentary that was allegedly killed for fear of the power of the diamond cartel can be found in Jani Roberts's evocative book, *Glitter and Greed: The Diamond Investigation.*[14]

As in combating illicit drugs, the United States has sent money to many of the destabilized African countries. Despite expenditure of a couple of billion dollars to support their governments, in excess of 10 billion dollars' worth of illegal diamonds have been smuggled out. And it should be noted that Americans buy 65 to 70 percent of all of the diamonds on the market annually.[15]

There has been recent response to the blood diamond crisis. The World Diamond Council, created at a meeting in Antwerp in July 2000, has admitted that human rights violations are of concern to them.[16] However, these efforts have been to shore up the illogical diamond market by increasing the penalties for selling diamonds outside the artificially inflated system.[17] The approach is addressed by Peter Meeus, head of the High Diamond Council, who states "Any trader that has dealt with these illicit diamonds will be banned out of the business."[18] Even the United Nations has condemned the blood diamond trade, while supporting the notion that the market can be controlled.[19]

The principles of this approach are fundamentally flawed. To keep the prices inflated De Beers must buy 60 to 75 percent of the world's diamonds.[20] In a world in which geological exploration is becoming increasingly sophisticated and diamonds are not really rare, the ability to control the market will be impossible. Sooner or later some diamond reserve holder will flood the market faster than the established cartel can respond. Their ability to buy the excess diamond production is already strained. When it gets to the bottom line, an industry built exclusively on image is unsustainable. Therefore, it is time to stop the charade and take the financial incentive away from those supporting terrorism.

Tomahawk missile sequence destroying ground target
Hughes Missile Systems

Tomahawk missile programmable warhead detonation
Hughes Missile Systems

Active Denial System—a millimeter wave antipersonnel weapon in research phase at Kirtland Air Force Base
John Alexander

Artist conception of the Active Denial System as it will be mounted on HMWWV. The ADS will cause pain at several hundred meters.
USMC

Seattle police using PepperBall and other non-lethal weapons during World Trade Organization riots
PepperBall Technologies

Joint Direct Attack Munitions Sequence
USAF—Air Armaments Center

Joint Direct Attack Munitions Sequence impact on target
USAF—Air Armaments Center

F-22 Raptor
USAF

Predator Unmanned Aerial
Vehicle—recently had
Hellfire missiles added
USAF

Global Hawk—Long Range UAV
USAF

New Virginia Class submarine—774
Virginia now under construction
U.S. Navy

Landing Craft Air Cushion
(SCAC) with full combat load
U.S. Navy

Members of SEAL Team Two train
with new SEAL Delivery Vehicle.
*U.S. Navy, Photographers Mate
First Class McKaskle*

Fireant Armed Robot detects moving target and prepares to attack a tank.
Sandin National Lab, Intelligent Systems and Robotics Center

Tank hit by Fireant
Sandin National Lab, Intelligent Systems and Robotics Center

New Commanche Attack Helicopter
U.S. Army Aviation and Missile Command

New Netfires system—known as "Missiles in a Box"
U.S. Army Aviation and Missile Command

Ranger firing Javalin missile at tank
U.S. Army Aviation and Missile Command

Javalin missile hitting tank
U.S. Army Aviation and Missile Command

Named Bird of Prey, after the Klingon spaceship, this recently revealed reconnaissance plane is designed for daylight stealth. It has low radar, visual, and acoustic signatures.
Boeing

Returning to the Roman phalanx, troops train to control rioters with non-lethal weapons.
USMC

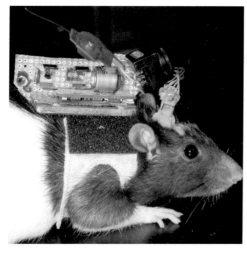

Sometimes known as RoboRat, researchers at SUNY Health Science Center implanted electrodes in the brain to direct the rat remotely. It will be used for searching areas inaccessible to other sensors.
Sanjiv Talwar, SUNY Health Science Center at Brooklyn

A member of Natick's Operational Forces Interface Group portrays the Objective Force Warrior. The OFW concept is a lightweight, fully integrated combat system for the individual soldier.
U.S. Army SBCCOM Natick Soldier Center

Soldiers training for very dangerous urban combat
U.S. Army

First crew arrival for inflatable manned station
Bigelow Aerospace

Voyager XM 300—inflatable manned space module
Bigelow Aerospace

STEP THREE

While illegal drugs are the single largest income producer for organized crime, it is the mechanism that moves money that makes everything work. When huge amounts of illegal cash are obtained, the problem becomes how to transition it into usable currency or trade goods. Hence the third critical step in eviscerating terrorism's financial base is to stop money laundering.

The term reputedly originated from Mafia ownership of Laundromats. To explain where they were obtaining money, which actually came from gambling, prostitution, and extortion, the likes of Al Capone bought legitimate Laundromats. These high-cash businesses were used to cover the illegal transactions. As Capone was convicted of federal tax evasion, Meyer Lansky sought new venues for Mafia money, including numbered accounts in Switzerland. He also developed the loan-back scheme in which illegal money was concealed as loans from compliant foreign banks.[21]

The amount of money being laundered is huge. Though the exact amount is not known, estimates range as high as over $2 trillion annually. A most conservative estimate by Michael Camdessus, former managing director of the International Monetary Fund, placed the annual amount at $600 billion.[22] Department of State officials confirm the connection between money laundering and drug trafficking, terrorism, arms sales, blackmailers, and even the burgeoning field of credit card swindles.[23]

Concealing money is not new. The practice can be traced back at least 3,000 years, to merchants of ancient China who sought to hide their wealth from the local rulers who wanted to place taxes on it. By the 1980s money laundering was growing rapidly. Advances in information technology coupled with globalization of finance and economics have made the process considerably easier. With just a few keystrokes, large sums of money can be moved electronically through financial institutions in several countries. This process makes the funds difficult to trace, and they often end up in countries that have difficult, even Byzantine, laws and procedures for enforcement.[24] The rapid movement of funds is not necessarily bad. In fact, it is a requirement to keep the global economy moving. However, organized crime has figured out how to disguise their transactions inside the ever-expanding international monetary interactions.

There are three basic steps in money laundering. First is placement, in which illegal funds are deposited in a bank or another financial institute. This may be done by direct deposits, wire transactions, or other means. The second step is called layering. Here the criminals move the money several times so that the connection between those placing the deposits and current ownership of the funds becomes blurred and difficult to trace. Third, there is the integration phase. At this point the illegal money is transferred into a legal business that provides cover for the origin of the funds. Then companies may withdraw funds from shell companies for nonexistent goods or services.[25]

One of the best-known money-laundering schemes was dubbed the Pizza Connection by the press. In the early 1980s organized crime elements bought a number of pizza parlors in midland America. They used the pizza restaurants to cover the distribution of a reported one and a half tons of pure heroin that took in $1.7 billion. The pizza business was chosen because it collects a lot of cash in relatively small amounts. The high volume of cash for pizza was recognized when Congress enacted the law requiring all deposits in excess of $10,000 in cash to be reported to federal authorities. Congress granted pizza and used car businesses exemptions from that reporting, making them ideal for Mafia operatives.

Nearly every country in the world is involved in financial crime at some level. In recent years there has been pressure to end secretive accounts and stress better international cooperation. However, for every effort to stem money laundering there have been loud claims of invasion of privacy of individuals, many of whom legitimately want to keep their financial dealings private.

Much is made of the offshore banks of the Bahamas and Cayman Islands. Those countries have tax laws that are not as sophisticated as those in the United States and Europe. However, electronic money frequently goes through State of Delaware corporations that are known to have minimal reporting requirements. In the United States, major banking institutions have been cited as not being sufficiently vigilant in looking for money laundering through their accounts. With approximately $350 billion floating outside the U.S. borders, the U.S. dollar has been the currency of choice. However, the strengthening euro will provide new opportunities for launderers.[26]

Global efforts are under way to counter money laundering. The Organization for Cooperation and Development created the Financial Action Task Force on Money Laundering (FATF) with thirty-one members. They did adopt a list of forty recommendations for countering money laundering. Unfortunately, the FATF has become highly politicized and focuses its efforts on nonmember countries.[27]

While some inroads have been made, money laundering remains a key issue for financing terrorism. What needs to be done in the United States is to tie foreign aid and other international economic assistance to efforts that help in tracking illegal money. In short, countries that receive assistance would in turn pass laws that make it easier to track financial dealings and allow prosecution of those caught money-laundering.

CHAPTER ELEVEN

POWER OF THE PRESS—A STRATEGIC WEAPON

THE FOURTH ESTATE HAS seen its power wax and wane. With advances in information technology various forms of news media have emerged. On the positive side has been the ability to acquire and disseminate information very rapidly. Using the Internet one can access dozens of newspapers and official government sites from around the world and elicit contrasting points of view. On the negative side has been the enormous amount of bad information that is put forth on a daily basis. Information has always been an important component of warfare. These new technologies have elevated its level of importance to critical status.

While tomes have been written about the use and abuse of the power of the press, manipulation of information, including that coming from traditional sources, has become an issue of strategic importance. While most Americans firmly believe that our First Amendment rights are among the most sacred, our adversaries have learned how to convert that great freedom into an effective weapon, one that can be turned against us.

The fact of the matter is that the news media have become strategic weapons. In future conflicts the ability to form and maintain international coalitions will be a critical issue. Similarly, generation of support of the American people will dictate whether any campaign will move forward domestically. Regardless of intervening events, the following example depicts the complexity and application of this nontraditional instrument of war.

At the time of this writing the outcome of efforts to overthrow Saddam Hussein is unclear. However, what is clear is that many parties are using the news media as strategic weapons. On a daily basis headlines around the world carry some aspect of the story. Influential leaders of all persuasions are quoted as to their opinions on the impending actions. Supporters of Iraq seek to separate those who might assist in U.S. efforts for an attack. Even Iran, an adversary of Iraq, warned they would side with their old enemy.[1]

181

Members of the Bush administration publicly outlined the case against Saddam and his motivation for acquiring weapons of mass destruction both on television and in press interviews. Russia, a traditional friend of Iraq, used the press to warn the United States against armed intervention.[2] The European media from countries with various motivations voice comments of concern about the legitimacy of an attack. Even Nelson Mandela, though not directly involved, made headlines decrying potential actions.[3] Later he was quoted as saying that it was *the United States that posed the greatest danger to world peace.*

The American press pushes the notion that they will serve as arbitrators between the Executive Branch and the American people in determining whether use of our troops is justified. The press exerts its influence by the frequency, timing, subject matter, and vocabulary used in these articles. Any astute reader can ascertain that the American media have gone far beyond reporting the news and are attempting to shape future actions. Though it is generally unstated, they, too, have a political agenda.

Closely coupled with use of information as a weapon is our inadequate educational system. One issue is the training provided those entering the news media profession. Fundamentally flawed is the notion that anyone who can write or talk well can automatically become a subject matter expert in any topic. Gone are the days in which reporters truly specialized in the issues that they covered. Instead, Lexus and Nexus can bring any reporter up to speed in a matter of hours. At least that is how many of them feel. Unfortunately, few have learned that no bad story, no matter how thoroughly refuted, is ever killed. The same misinformation that fed the original story is omnipresent, and it is a matter of how an Internet search engine prioritizes information that determines what the investigator will see and in what order. Often under intense pressure to get the story out reporters rarely get past the first seemingly articulate entry. Without their own expertise on which to rely, they frequently choose poor sources, often on name recognition. Even prestigious institutions and individuals get the facts wrong.

Another aspect of poor education pertains to those who receive information. Unfortunately, many Americans have little, if any, sense of history or geography. We are lucky when students can identify states in the middle of America, let alone foreign countries. When Saddam Hussein, or at least some Iraqi official speaking for him, states in an E-mail to Christopher Love: "Hence I remind you of the Crusades in 1096–1291 . . ." we can't even fathom what is being discussed.[4] Effortlessly Saddam reaches back further in time and develops the concept of brotherhood from Adam and Eve, acknowledges Abraham "the father of all prophets," and then binds these thoughts with the impact of colonialism.

In the United States we simply do not comprehend the importance of history and consistently focus on near-term solutions based on symptoms. Rarely do we attempt to understand causal relationships. If the problem can't be stated in a twenty-second sound bite, we just forget about it. Simple slogans gain inordinate power whether they make sense or not. They form the basis of opinion polls. Opinion polls are followed daily on Capitol Hill. Votes frequently are cast accordingly, especially if an election is approaching. While proclaiming the wisdom of the American populace, politicians well understand the need to dumb down their messages. The unfortunate reality is that some of them are not much better educated on current topics than their constituents.

Far worse will be the iconoclastic specter that oozes up from the primordial depths of misinformed interpretation of Islamic thought. With secular education virtually annihilated in many Middle Eastern countries, uneducated but articulate and persuasive imams have fertile mental fields to plow. And so they go about their devious ways sowing seeds of discontent, brainwashing young minds, and creating new cohorts of terrorists—ones who gladly throw themselves onto the funeral pyres of history. Their real reward is having their picture on a trading card that will inspire the next generation of misfortunate children to follow in their tragic footsteps.

It is not education alone that inspires the followers of jihad. As learned in the aftermath of 9/11, the terrorists who flew the planes were educated men. It is the ability of the influential religio-political leaders to co-opt the thoughts of millions of people and derive willing obedience to acts unthinkable to most Westerners. With aplomb these leaders warn us of an inexhaustible supply of suicide bombers

Previously lacking in multimedia skills, the imams have quickly grasped the importance of public relations and mass communications. They stay ahead of the West in most instances in presenting their side of the story to the people. Adroitly the imams turn nearly every incident to their favor. While their people receive only carefully scripted messages, these leaders have been successful in presenting their story to the more open Western media. The degree to which they have been successful can be seen demonstrated in the claims by Congresswoman Cynthia McKinney (Democrat, Georgia), who supported the notion that President Bush and other officials had advance warning of the 9/11 attack but did nothing to stop it.[5]

In short, the Western press has little impact on the burgeoning masses of the Middle East. However, the crafters of Islamic public sentiment have been gaining ground steadily. They respond rapidly to current events and effectively spin stories for both internal and external consumption. To counter this we will need a four-dimensional military as outlined by Chuck de Caro, one that understands and operates with ease in the information war.[6]

THE SOURCE OF OUR KNOWLEDGE – LOOKING GOOD

Anyone who has watched television evolve over the past decade cannot have missed the ubiquitous emphasis on beauty and entertainment. A few of the old guard commentators remain, but they are a dying breed. Morley Safer, for instance, would not rate high on the Q factor for the coveted young demographic audience, but the point is made. The real casualty in the endless race for higher ratings through more physically attractive hosts has been experience. When interviewed on Fox News Network by Linda Vester, I was pleasantly surprised that she was a Fulbright Scholar and specialized in Middle East affairs. Bill O'Reilly, no matter what you think of his politics, has paid his dues through years of foreign travel. Larry King has gained a global reputation for his insightful questions and ability to draw key people. With a few others they are the exceptions. Many reporters and commentators are relatively bright, articulate people who rarely have a clue about the topics on which they are speaking. But there are too many of them attempting to influence us who have IQs just above that of a rock.

What the young beauties of the cable networks fail to realize is that they are but an intermediate point in the transition. The ultimate in combining programming and entertainment, *Totally Naked News*, probably was only slightly ahead of its time. Today live reporters must worry about the next set of younger and cuter talking heads to come along. However, the real threat to their longevity is technology. Soon there will be no need for them at all. Computer programming is sufficiently advanced that we will soon see artificial computer-generated imagery (CGI) news reporters that are specifically designed to meet the desires and expectations of the audience. In fact, you will probably be able to select the CGI reporter that most fits with your fantasy, although the news content will be the same from any visual aid available.

Digital actors have been a concern in Hollywood for some time. Actors of the human variety understand that as the CGI technology improves, their services will be needed less frequently. In fact, the recent New Line movie

Simone portrays Al Pacino creating a movie star based on a near-mystical CGI system that is probably closer to reality than they would like to think.[7] The difference is that Pacino presents Simone, which is actually a contraction of Simulation One, as a human. In the highly competitive special effects field, Industrial Light and Magic may lead, but others are always close behind.

While this technology is expensive at present, those costs will go down and quality will go up. Better yet for studios, they will not have to pay exorbitant sums to prima donnas and put up with their exotic idiosyncrasies. Studios may choose to have a human who can handle live appearances for them, but the salaries will be modest and new models can be introduced at will. No longer would a studio have to conduct damage control when stars engage in scandalous or illegal activity.

Employing CGI for news personalities may actually be a bit trickier. The physical appearance will not be a problem. Rather, it will be the degree of trust and confidence that is conveyed to the reporter. Of course, this notion of personal reliability has been in steep decline in recent years, as audiences have become more cynical about news reporting. Gone are the times when Walter Cronkite was one of the most respected public figures in America. As books exposed the inner workings of the network news, people became more concerned about the veracity of the material that is synthesized by faceless curmudgeons and force-fed to the public.

APPLICATIONS

Of greater concern should be an understanding of the motives of these molders of public opinion through manipulation of the media. The new CGI will offer another step in insulating those providing information from the audience. We have already seen how synthesis, spin, and selection of stories to be covered have significantly influenced public perceptions. Now we add another degree of separation, but at our own peril.

It is clear that our potential adversaries understand the importance of perception management. Public relations is no longer the sole domain of countries or large organizations with sophisticated staffs. With computers and fax machines in every corner of the world, any savvy operator can gain access to the outside world and present his case to a world that does not discriminate between credible and non-credible sources.

During the conflict in the Balkans it was not uncommon for faxes to be sent from cities under siege. The senders had access to the fax numbers for news agencies and groups likely to be sympathetic to their cause. Concurrently messages sent on the Internet proliferate at phenomenal speed. I have received messages from unknown sources stating that they are under attack at that moment and asking for help in getting the U.S. government to intervene on their behalf. Frequently there is a personal appeal or a linkage established between the sender and receiver. A common ploy is to state that the sender is an American student who returned home only to become trapped in a desperate situation.

However, groups such as al Qaeda are very sophisticated in their material designed for public consumption. The recruiting tape they sent out was extremely well crafted and tailored for their target audience. Even the tape that graphically detailed the execution of Daniel Pearl hit its mark and was enthusiastically received in some sectors. While most people in the Western world were shocked at the notion that anyone would videotape and display the decapitation of an innocent civilian, the act was cheered in some areas of the world. The messages sent from Osama bin Laden following the 9/11 attacks were full of symbology and well crafted for the disenfranchised youth of fundamentalist Islamic countries.

Domestic press releases have evolved to a level of concern. During the summer of 2002 a series of articles appeared in major papers addressing plans for attacking Iraq. These articles purported to have obtained inside information about plans the Pentagon was considering as means for removing Saddam Hussein from power. Included in these discussions was material about when and how attacks would occur. Part of the material related the countries that would support the United States in launching an air and ground attack. The possible movement of special operations forces and the groups with which they might collaborate were detailed.

The response from Secretary of Defense Rumsfeld was direct, as he noted that any such leaks were totally unauthorized. Pentagon sources did state that various contingency plans are always drawn up for areas of interest. This is known as being prepared and does not mean that any of the plans will be executed.

There is a remote possibility that such plans were intentionally provided to the press. After the stories broke, we learned that Saddam put Iraq on a wartime footing. In so doing he began adding stress to the Iraqi military. It is difficult to maintain a high degree of preparedness for a long period of time. Such indecisiveness we, too, have experienced with the multitude of nondescript terrorist warnings that were put out by federal agencies after

9/11. Imagine the anxiety raised by constantly being warned that bombs and missiles were coming. Therefore, one explanation could be intentionally getting the adversary to overreact.

There are two more plausible explanations. One is to preempt the authority of the Executive Branch by leaking real plans. In so doing those providing the information can shape the actions of the military simply by describing operations they do not want to occur. Deception is a huge part of military operations. Since releasing real plans could cause higher casualties, airing options deemed undesirable by any party could prevent a plan from being executed.

Of course the third alternative is just plain poor security. It is known in Washington that if two people know something it is not a secret. There are many reasons for leaking official secrets. None of them is acceptable. The press, in general, has assumed an adversarial position with the Government. While there have been occasions on which the press embargoed a story, they are few and far between. The constant rejoinder is the public's right to know. To quote Tom Clancy in a television interview about activities at the super-secret Area 51, *that doesn't mean the public's right to know everything.*

The media's role of self-proclaimed protector of the people must be challenged. Somewhere the notion got started that the press is immune to prosecution for misdeeds done in the name of the public good. Reporters repeatedly demonstrate the blinding flash of the obvious by penetrating security systems such as those at airports. Then with fanfare they announce what everyone always knew, as if they have made a monumental discovery. One such venture sponsored by ABC included moving radioactive-depleted uranium across Europe and into New York without detection. Of course, it was really part of a gimmick for their 9/11 programming.[8]

At about the same time, Sky News in the United Kingdom assigned Ms. Islim Volcan to attempt to get a job in the Frankfurt airport that would allow her access to British Airways aircraft. They provided her with an obsolete and reported stolen identity card, which she successfully used to obtain employment. She then secreted a cell phone with simulated plastic explosives on a plane where an accomplice making the flight retrieved it. As I saw the program aired in Manchester, there can be no doubt that Ms. Volcan committed a series of illegal terrorist acts. Again there had been complicity by senior executives of her company.[9]

Authorities should ensure there is no get out of jail free card. In fact, if apprehended, such reporters should be tried as terrorists and sentenced to similar prison terms. Concomitantly, the senior personnel who authorized the escapade should be arrested and tried on conspiracy charges. One conviction should end the practice.

Unfortunately, these same ersatz defenders of truth have also become the vigilantes of modern America. Repeatedly they have attacked innocent people without hesitation or mercy. Once an individual has been identified as a suspect, the media begin an unrelenting feeding frenzy, each outlet attempting to beat its competitors. Frequently they bring in experts with no direct knowledge of the situation and ask them hypothetical questions that strongly imply the guilt of the suspect. Television and Internet media create endless polls asking viewers to express their opinions about people and issues, even though the facts are not available.

Anyone doubting the validity of the shark mentality should review the case of Richard Jewel. He was the person who reported the bomb at the Olympics in Atlanta and became a suspect in the case. Years after the reporters' vicious attack, which included staking out his mother, his life was basically in ruins and the media defended their actions. The media lawyer's belief that prefacing any statement with the word *alleged* justifies it and relieves the outlet of all responsibility is nonsense. At a time when the media play a critical role in determining public support on key issues, inaccurate or irresponsible reporting can have significant consequences on national security.

In addition to the news media, other elements of that industry are having a significant impact on our national security. There has long been controversy about the roles and responsibilities of entertainment providers. Inside the country debate rages about the potentially deleterious effects of rap music and violent crime as depicted on television and in movies. Defenders of First Amendment rights believe that the effects are minimal and the greater good is served by allowing free expression. They argue, albeit unrealistically, that parents will monitor what young children watch. Further, it is suggested that we are sophisticated viewers and can distinguish between entertainment and reality.

However, a recent survey indicates that this programming is having serious consequences in unanticipated ways. When Professors Margaret and Melvin DeFleur of Boston University conducted a study of the attitudes of teens toward the United States in twelve countries they discovered that these teenagers felt very negatively about the United States and that for the most part their perceptions were developed through television programs. In fact, fewer than 5 percent of the people surveyed had ever met an American.[10]

While such negative perceptions about our morality and virtues might be anticipated in Saudi Arabia, Bahrain, and other Islamic countries, many of the others were shocking. For instance, views in South Korea and Mexico were nearly as negative as those in Islamic states and Taiwan. It should be a matter of concern that this is despite the fact that the United States has guaranteed the security of Taiwan and South Korea for decades. The positive effects of our enormous security and economic contributions to these nations have been eradicated by our profit-driven and unconstrained entertainment industry. This does not bode well for future coalitions and other international relationships. The continued development of these attitudes will assist in recruitment of the next generation of terrorists. As with the news media, those in entertainment are rarely held accountable for the impact of their actions. There is a great failing in our educational system. They do not teach that actions have consequences—and sometimes these are unintended. In the former case, these consequences were foreseeable.

ERGOFUSION

I would like to introduce the concept of ergofusion. Defined as "the misidentification of causal relationships," ergofusion is prevalent in the fields of non-lethal weapons and combat in general.[11] Examples abound and are derived intentionally when twisted statements are made to prove a point, evoke a response, or mislead the reader. Ergofusion also happens inadvertently due to lack of critical thinking skills. In general, our educational system fails to teach deductive logic adequately, leaving many people ill prepared to decipher the ubiquitous spin.

Non-lethal weapons have been frequent targets of ergofused concepts. For example, it is asserted that the availability of these weapons will necessarily lead to increased propensity for conflict or more pain and suffering. That basic thought goes, if we have non-lethal weapons we will be more likely to use them, and disregards any notion of conscientious decision making on the part of legislators.

An example occurs when any form of gas is used against citizens. Even a relatively benign substance such as pepper spray or tear gas will often evoke the media to equate that application to the use of Zyclon-B, the lethal gas in the Nazi death camps. Obviously, there is a significant difference, but this extrapolation plays to the emotional value of the argument. This technique has been seen in papers serving communities with large Jewish populations.

The reality is that technologies do not cause bad behavior. It is the people who use technologies for evil purposes that demonstrate bad behavior. Blaming non-lethal weapons for increased conflict or human suffering is a quintessential example of ergofusion.

The concept extends to many other areas. It is quite typical for bureaucrats to confuse "I can't" with "I won't." The difference is causal relationships. "I can't" implies a physical blockage that prevents the action from occurring. "I won't" means the action is possible, but for some reason the bureaucrat has decided against doing it.

A defense-related example of ergofusion happened on the late-night news/satire program on Comedy Central called *The Daily Show*. Jon Stewart was interviewing Sen. Charles Schumer of New York shortly after the *New York Times* ran the stories about strategies for attacking Iraq. Stewart noted emphatically that this was true democracy and that the American people had a right to participate in these decisions. The articles discussed where and how the attacks could occur. Those are military decisions and leaking such information is treason. Discussing whether an attack should take place would be part of the democratic process. Stewart was practicing ergofusion.

A larger example of ergofusion that is central to counterterrorism is the notion that cursory screening at airports makes it safer to fly. In reality, the most dangerous man in all transportation security is Norman Mineta, the Secretary of Transportation. It is under his watch that we have seen the creation of the Transportation Security Administration and implementation of the incompetent random screening process. Under that system all passengers, no matter their demographic status, have an equal chance to receive in-depth scrutiny.

Mineta, because of his early childhood experiences, refuses to permit the use of profiling to identify potential terrorists. This act of reckless endangerment places the lives of all passengers at risk. Profiling is the single best method to spot terrorists before they get on an aircraft. To date, not a single seventy-year-old grandmother has bombed a plane. Yet they are equally subject to the searches. As an example, a personal female friend who is close to that age was selected randomly on all three legs of the same trip for extensive searches. This is nonsense and should have been grounds for removal of the person establishing such a policy.

The fundamental logic flaw in the screening process is confusing randomness with probability. Subjecting the very young, old, and everyone in between to the same likelihood of more intensive searches invokes randomness. If terrorists were determined to be from the general population and no one could be trusted, that would be valid. They aren't. The fact is that most suicidal terrorists come from a relatively well defined demographic group. The probability that young males from Middle Eastern countries are terrorists is orders of magnitude higher than the probability of any other subset of people in the world. Next highest on the list would be females from certain Middle Eastern countries. The next level to watch for is the unwitting dupes who have been in close contact with the former groups.

Interviewed on *60 Minutes II*, Mineta was asked whether a person with the name Mohamed should have a higher chance of being searched than anyone else. He responded, "I hope not."[12] The same held true, he thought, for males from the Middle East between the ages of eighteen and thirty-five. Even though most of the highjackers on 9/11 fit that profile, Mineta willfully submits all of us to the same level of scrutiny due to his misguided, anachronistic politically correct beliefs. As effectively demonstrated by the Israeli airline, El Al, profiling is the single best tool for detecting potential highjackers.[13] Considering there are limited resources available for airline security, it makes sense to concentrate them on areas of the most likely danger. Mineta commits ergofusion when he confuses being politically correct with our security and random searches with probability. The three best methods to determine the likelihood that an active terrorist is attempting to board a flight are profiling, profiling, and profiling.

As an aside, airline security while in the air is a simple matter—deterrence. First, arm the pilots with lethal weapons. Next, provide flight attendants non-lethal options such as the M-26 TASER. Finally, provide every able-bodied male, and willing females, with non-lethal plastic truncheons. In that way any potential highjackers will know they will probably be beaten to death if they attempt a terrorist attack. The rules of passenger pacificity changed on 9/11. As has been repeatedly demonstrated, no screening process will ever provide 100 percent assurance for preventing weapons from getting on board.

Apparently neither Mineta nor anyone else in the federal transportation security system ever watched *MacGyver*. Anything can be used as a weapon. For example, credit cards can quickly be sharpened into cutting instruments. Therefore, terrorists should know—not guess—that adequate force to thwart any attempt to highjack a plane is always available.

HISTORY CHANNEL EFFECT

A new issue has been emerging over the past few years. I call it *the History Channel effect*. Of course, it is well established that the winner writes the history of conflicts. However, we are currently witnessing cathartic urges to produce revisionist versions of prior events and expose any perceived misdeed in the most egregious light. While it is argued that such presentations make for a better-informed public, too frequently they are really stepping-stones for overzealous reporters. Like all critics, they come well prepared to tell others how they should have conducted their affairs. Unfortunately, they rarely possess the skills or courage to perform the combat operations on which they feel perfectly qualified to comment with the clarity of twenty-twenty hindsight.

The few who do report accurately themselves become targets for criticism by these armchair generals. Take, for instance, Joe Galloway, who was the only reporter present at the pivotal November 1965 battle for the Ia Drang Valley in Vietnam. Later he and Lt. Gen. Hal Moore, the battalion commander at the fight, went on to write the universally acclaimed book *We Were Soldiers Once . . . and Young.* The battle was vicious, up close and personal. For his actions under fire Galloway became the only reporter in Vietnam to be awarded the Bronze Star for Valor by the U.S. Army.

The movie adaptation of the book, starring Mel Gibson as Lt. Col. Hal Moore, was criticized by David Denby in the March 2002 issue of the *New Yorker* for being revisionist. In Denby's view, this real-life depiction did not meet the standards he had acquired from contextually muddled films like *The Deer Hunter* and *Platoon* or the total nonsense of *Apocalypse Now,* all produced with a blatant political agenda. In reality, seeing the aforementioned movie with the accurate depiction of the characters by Mel Gibson and Sam Elliott, playing Sgt. Maj. Basil Plumley, is as close to combat as most reporters are ever going to get.[14]

The propensity for invoking the History Channel effect will further complicate future conflicts. Commanders will feel a necessity to dutifully record every aspect of the actions to cover their rear. This will have similar impact to the well-known *CNN effect,* which is now commonly accepted by both the media and the military. This refers, of course, to the cameraman who seems to shadow every squad and reports combat events to the world in real time.

The military understands the implications of the CNN effect and generally lacks trust in the integrity of reporters. Therefore, they have taken additional steps to distance reporters from ongoing operations. These include establishing pools and controlling access to U.S. units on combat missions. Another innovation has been restricting the use of last names of service members. Ostensibly to protect the soldier's identity and that of family members, it also separates them from the public who will provide support more readily when they can fully empathize with that person who is in danger. Lacking access to current events, save for sanitized versions from public affairs officers, reporters are more enticed to conduct retrospective examinations of battles. With the outcome known, the only avenue for aggrandizement is to ferret out previously unreported titillating information—the nastier, the better.

The History Channel effect may be inevitable, but it comes with a heavy price when revisionists contort events to their own agendas. There are many factors that exacerbate these stories. Lack of combat experience by reporters leads the list. Today few reporters have actually engaged in combat, yet they all think they have the innate sense to provide the truth. In reality truth is a scarce commodity and is generally in the eye of the beholder. As movie producer Joe Hyams once told me, facts change. In combat they change more frequently and there is a reason that soldiers talk of *the fog of war*. Yes, evil does exist and pure evil is easily discernible. But combat is usually about men engaged in doing the bidding of others. Rarely do the fighters espouse philosophical thoughts. For troops locked in a fray, the most important issue is survival.

War is too often the choice of young reporters looking for the fast track to fame. Unfortunately, their academic preparation rarely includes *Life 101*, and they are woefully ill prepared to attempt to make sense of the incongruities and paradoxes inevitable in war. Today as we send off our young to fight, most reporters come from a generation accustomed to convenience and lacking in the philosophical depth necessary for perspective.

High emotional impact is a critical factor in getting a report published or aired. Members of the news media are selling products in a highly competitive environment. Emotion brings in higher ratings. Sex and violence sell and war offers a cornucopia from which to draw. Frequent reports of rape come to the fore while other crimes, with even greater impact, go virtually unnoticed.

Historical stories often suffer from poor or selective documentation by reporters. Beating deadlines is endemic, and rather than spending time researching, reporters either use the reports provided by others or select the facts that support their position. Even in reporting past events, there is tremendous pressure to file as quickly as possible. With little personal experience to draw from, reporters are faced with overwhelming mountains of unevaluated information. Determining what is reliable can be a daunting task.

In review of older events we should be concerned about revised standards of conduct. During the Clinton administration there was a strong push to right past moral indignities. Among the leaders in exposing past errors was Hazel O'Leary, then Secretary of Energy, a position for which she was proven unqualified. At Los Alamos we and other national laboratories were pressured to release data about various experiments that had taken place in the early days of the Cold War. Some of these included human subjects, with and without their consent. In interviews of the participants, they invariably began by stating, "It was a different time." That was not meant as an excuse but referred to the deep-seated fears that were endemic in that period of global insecurity. Under those circumstances decisions were made that seemed to be acceptable at the time and by their standards.

In 1994 I was privileged to be tasked to look at the original records pertaining to the decision to drop the atomic bomb on Japan. From the comfortable distance of fifty years, many socially conscious people questioned why we could not have conducted a demonstration off the coast, instead of devastating Hiroshima and Nagasaki. Contrary to the assertions of some people, considerable thought and debate went into that decision. Many factors were weighed. One was that only two bombs were immediately available. Having tested only one device in New Mexico, and that of a different design, the scientists of the Manhattan Project could not state with certainty that the next bomb would detonate. Attracting attention to a failed experiment would have been disastrous.

The war in Europe was winding down, and casualties in the Pacific were high. The Normandy invasion demonstrated the problems of a very large-scale amphibious assault. It was estimated that the invasion of the Japanese mainland would be at least ten times as difficult as Operation Overlord. In prior island attacks the Japanese defenders had proven themselves to be courageous, resourceful, and sometimes fanatical fighters. In defending their homeland it was estimated that most would fight to the death rather than surrender. Rarely mentioned is that Operation Coronet, the actual invasion plan for Japan, called for the use of nine atomic bombs.

For those who still doubt the appropriateness of the decision, I recommend they travel to Los Alamos and read the guest book at the museum. See the thanks of the men who were en route for the dreaded invasion. Read the comments of their children who would not have been born had the bomb not been dropped. With these events the History Channel effect has certainly blurred all reason and we have repeatedly experienced reevaluation of historical events based on today's standards.

Several examples of media reports serve to illustrate my History Channel effect point:

Tail Wind. Based on an inaccurate report, CNN claimed that SF soldiers targeted U.S. servicemen believed to be assisting the North Vietnamese Army and employed lethal nerve gas against them during the Vietnam War. The incidents didn't happen, but that did not stop a highly publicized segment on the June 7, 1998, *NewsStand* program and several controversial articles from appearing. Unlike the case with most inaccurate stories, the producer, April Oliver, and her boss, Jack Smith, actually lost their jobs. Both Ted Turner and CNN president Tom Johnson made rare public apologies. Still, considerable damage had been done to the credibility of everyone involved with this story.

The Battle of Rumaila. In the May 22, 2000, issue of the *New Yorker,* Seymour Hersh spun a tale about this battle that took place in the waning moments of the Gulf War. Based on erroneous reporting of events from Desert Storm, Gen. Barry McCaffrey was portrayed as a war criminal and personally castigated for having violated the rules of war. As was well known to Hersh, the alleged incidents had been thoroughly investigated at the end of the war. McCaffrey was exonerated. The most probable reason for the story reaching print was that General McCaffrey had been promoted twice and was then appointed as the Drug Czar. Still, this inept reporter, knowing full well of the results of the investigation, was able to cast aspersions on the reputation of an American hero. The purpose, no doubt, was a vain attempt at personal glorification of a would-be investigative reporter—all at the expense of the U.S. military.[15]

The Hurricane. This was a popular movie based on falsehoods and a very slanted character portrayal of a convicted killer who had his sentence overturned. Important evidence in this case, conveniently omitted in the movie, indicates that Rubin Carter probably was responsible for the murders.[16] Yet Hollywood, with a moving performance by Denzel Washington, transformed Carter into a folk hero.

After Enduring Freedom, reporting got even worse. Neophyte reporters searching for any story that would raise their visibility came across tales of atrocities that may or may not have occurred. Apparently uneducated in the ways of Afghanistan, they allegedly were shocked to hear reports of limited atrocities committed by warring clans. Had they bothered to read any of the accounts emanating from the former Soviet Army when they fought in Afghanistan, the reporters would have known that harsh treatment of captives was the norm. While no American involvement was alleged, reporters began looking for the nearest troops who might have been knowledgeable about the atrocities. The reporters' intent clearly was to assess blame for even being aware of the misdeeds of others.

Frankly, in war things happen that most people really don't want to know about. Winning wars frequently establishes paradoxes and requires abridgement of values. War is not the fastidious noble clash as often portrayed in books and movies. The emotional rush, and later mental anguish that follows, when you look your adversary in the eye and kill him, cannot be conveyed by any medium of reporting. The notion that a better-informed public will somehow gain understanding from vicarious visceral experiences is nonsense. In the current state of human affairs war remains inevitable. Therefore, my admonition to those who would report on war is to first engage as a participant, then reveal your deeds before evaluating those of others.

In recent times the media has attempted to set itself apart from the society from which it was spawned. After 9/11 there were a few notable exceptions in which reporters openly stated they could not remain unemotionally detached. In fact, the media's self-serving objectiveness has been used to excuse a host of sins. Foremost among them is a lack of responsibility and accountability. In feeding frenzies fostered by personal competition and unreasonable time constraints reporters willingly and unapologetically destroy lives and institutions. Errors are rarely acknowledged and never repaired. In the truest meaning of the words, the media have become weapons of mass destruction. The paradox that faces them today is whether they can exert sufficient temperance so that they do not enervate the very society that has fostered them, thus contributing to their own demise.

CHAPTER TWELVE

THE EPITOME OF PRECISION

DURING THE PAST THREE decades the U.S. military has spent billions of dollars on the development of precision-guided weapons. From both short and long ranges they are programmed to hit only their designated targets. In Desert Storm there were extensive images on television of laser-guided bombs hitting an illuminated building. We watched as cruise missiles, using terrain-following guidance systems, flew through the streets of Baghdad to their target. While these weapons were not perfect, most military analysts believed the high cost of development was easily offset by the reduced number of bombs needed to destroy a specific target. As noted by Col. John Warden, who drafted the air campaign for that battle, the Air Force destroyed more key targets in one day than was accomplished in any year during World War II.[1]

Since Desert Storm, additional money has been put into research and development of all-weather precision munitions capabilities. This effort was in response to the cases during the Gulf War in which intermittent cloud cover caused the laser to lose track of the target, allowing the expensive ordnance to go astray. As weather in many areas of the world is less than optimal for long periods of time, the ability to direct bombs through clouds was critical. Weapons also were improved so that as missiles were in flight they could receive updated information about the target or be redirected to a newly discovered target of higher priority.

Such precision-guided weapons are needed for future conflicts. But they belie the truth about the political aspects of their use. The epitome of precision is assassination! In recent decades the United States has generally accepted decrees that we would not be involved in assassination, especially of foreign leaders. During Desert Storm great efforts were made to determine the exact location of Saddam Hussein at any given time. Even then, there were euphemisms provided the press so that we did not admit that he was the specific target. It was stated that we were attacking the military leadership of Iraq—which of course was led by Saddam. Such statements also provide wiggle room when the attacks fail to take out the targeted individual. As with Saddam, Osama bin Laden, and others, individuals on the run have often proved difficult to locate and eliminate.

In discussing the emotional resistance to assassination with counterterrorism expert and former Delta Force operative Eric Haney, he pointed out that the word has assumed a connotation as an *unwarranted and illegal killing of a specific prominent and high-level figure.*[2] Instead, assassination may be viewed as the targeting for death of those combatants and their supporters who have demonstrated a willingness to employ terrorism against us—including our civilian population.

At one of the early conclaves on non-lethal weapons, the assembly addressed the paradox created between precision and assassination. The issue arose during a discussion about whether it would be better to fund non-lethal weapons or improve precision. Precision, it was claimed, decreased collateral casualties. There was consensus that assassination was a viable alternative to both very destructive precision munitions and massing of fires on urban targets. It was decided, however, that if the committee mentioned being in favor of assassination in the published report it would totally obfuscate the main focus of the study—gaining support for non-lethal weapons.

Assassination is clearly an issue in which emotion overrides reason. Politicians are perfectly willing to kill thousands of soldiers and accept both friendly and collateral casualties with bombs and bullets marked "To whom it may concern." Rarely are they willing to sign an order that states: "Here is a bullet for Osama," or any other despot for that matter. For those who order young soldiers into combat, adding that final degree of personality becomes distasteful. Unfortunately, these policies do not provide the same psychological distancing for the troops who join the battle, pull the triggers, and come face-to-face with the physical effects our highly lethal systems deliver. For it is the infantrymen who encounter bodies torn asunder, smell the stench of decaying human matter, and hear the screams of agony of those soon to die. They know too well the awful effects war inflicts on non-combatants, young and old, who, trapped in areas locked in mortal strife, become unwilling pawns—and too frequently dead. All this because politicians falter egregiously and fail to accept that a single bullet could avoid the carnage and change the course of history.

Now the War on Terror has brought a new focus on assassination as an implement of conflict.[3] Though assassination is used extensively in some parts of the world, the United States has been quite reticent to expand the policy severely limiting the practice. That will change. It is only a matter of time, as terrorists employ their dastardly acts and we suffer sufficient casualties to know we must adopt a policy that is ruthless when dealing with them. As appropriately stated by military author Ralph Peters, there is only one way to deal with terrorists—you kill them.[4]

Despite our unparalleled military might, many of our adversaries view us as weak because of our unwillingness to act when we have been attacked. For all the effectiveness of Enduring Freedom, our response was restrained. Partially that was done to establish and maintain alliances. They are fleeting and self-serving and in the end will leave us more vulnerable than before.

There are two main counterarguments to assassination. One suggests that the oppressor you remove may be better than the one who replaces him. That may well be the case, but you will know one is gone and the next put on notice. The second position deals with maintaining the moral high ground. It is simply not nice to go around killing selected adversaries. Partially this is self-preservation for our political leaders. They do not wish to raise the probability that they may become targets. Since we claim to be a nation of laws, there is the notion that anyone deserves a fair trial before being imprisoned or executed.

However, several of our self-righteous allies will not extradite persons who could be tried for capital crimes—including terrorists. Sometimes these allies will not even provide evidence to support such cases. Even allowing the circumstances that led to such laws and constitutional considerations in Europe, to withhold evidence in a case such as that of Zacarias Moussaoui, the alleged twentieth highjacker, seems beyond the pale.[5]

Fighting terrorists will be very, very different from conventional wars. Strict adherence to the current laws of land warfare will not serve us well. In similarity to many treaties, these laws were designed to serve another function. They assume that national armies will meet on a great battlefield and fight until one is declared the victor. Both sides would take prisoners and care for casualties regardless of sworn allegiance. This was indeed an extension of diplomacy by other means, albeit one with finite limits. In the last century the major wars lasted from days to a few years at most. In the end repatriation was the norm, modestly friendly relations were reestablished, and the winner paid the bills.

Terrorism alters the boundaries. Asymmetric in nature and nearly timeless in scope, these battles will ebb and flow for decades. The status of prisoners begs questioning: How will we handle those detained caught perpetrating acts of terrorism? In combat there is no difference between soldiers who shoot and support troops. Once captured, they were all prisoners of war. Now even differentiating between crimes and acts of terrorism, or those that enable terrorism, will be nearly impossible. Will incarceration be indefinite, possibly for many decades, without judicial proceedings? Typically POWs have a designated status, but certainly trials and a prescribed length of sentencing are not among their traditional rights.

The quandary has begun. Those detained in Afghanistan have fallen into various categories. Local tribesman simply took an oath to the new leaders and went home. Foreign combatants and senior Taliban officials were detained in several manners. Selected al Qaeda members were detained in country while others were transported to Guantánamo, Cuba, for processing.

The al Qaeda terrorists claimed to be of many nationalities, Saudi, Pakistani, Egyptian, Chinese, and Chechen, and included Americans. Various countries began independent negotiations for repatriation of their people. Some held by Afghanistan without U.S. oversight were returned to Pakistan directly. This despite the fact that the prisoners openly stated they would return to the *jihad* and again fight against Americans and other enemies as soon as they had the chance. Clearly the interests of the Afghans and Americans were already at odds.

John Walker Lindh, the American Taliban, was taken into custody during combat in which Americans died and yet was brought into U.S. federal court. While he plea-bargained and accepted a prescribed sentence, during pretrial maneuvering his lawyers questioned whether he ever committed acts of combat against Americans. While the question should not be considered relevant, it does raise flags about legal strategies for future detainees.

Unlike conscripted warriors for states of old, these terrorists are highly motivated and determined to accomplish their objective at any cost. Since dying for their cause is not only acceptable but also desirable, they cannot be trusted to sit on the sidelines once they have been captured or repatriated. They will not stop until they succeed in destroying our way of life or die trying. That presents a legal, moral, and philosophical dilemma. If we are to survive the coming war we will have to make some very hard choices, ones that will run counter to Marquis of Queensbury rules of etiquette for warfighting.

It begins with Peters's solution. Being politically incorrect, I'll run up the Black Flag—no quarter asked, no quarter given. First we need to kill as many terrorists as possible without allowing them to surrender. The next step is even harsher and is currently unacceptable. It will not be considered until sufficient pain has been inflicted on Americans that they are prepared to do whatever is necessary to exterminate the threat. It is also fraught with danger. Suicide bombers expect that they will go to Paradise. They also anticipate that their families will be adequately taken care of after they complete their mission. It is proven that Palestinian terrorist support groups, including some under Yasser Arafāt's control, pay substantial amounts of money to the families of suicide bombers. It is also known that members of the Saudi royal family and other countries have made large contributions to help the survivors. Claiming they are only helping widows and their children makes as much sense as the boy who kills his parents begging the court for mercy because he is an orphan.

The currently unthinkable action step is to put the families of terrorists at risk. When suicide bombers are identified all known family members should be targeted for assassination. However harsh, this is the only method that has a chance of stopping the attacks. The one thing that the suicide attackers wish for is protection for their families. So far that has been granted with cash payments. The bombers may believe they will get points in heaven for blowing themselves up, but they don't get any for being responsible for wiping out their own relatives.

There is an urgent need to be able to counter suicide bombers. With explosives and a brain they are smart weapons that have pinpoint accuracy and can determine the best time to execute their mission. During 2001 and 2002 they accounted for about 40 percent of all deaths in Israel from Palestinian attacks. More important, they have significantly altered the casualty ratio. Prior to September 2000 the ratio was about 6.7 Palestinians killed for every Israeli citizen who died. After attacks resumed in March 2001, suicide bombers changed the ratio to 1.7 to one. Also of concern is the new demographics of the bombers. They are no longer restricted to militant young, single males, as older, sometimes married men and a few women have joined these ranks.[6] As a few others have predicted, I agree that sooner or later we will experience suicide bombers in U.S. shopping malls or other public gathering places.

Israel has used halfway measures such as destroying the homes of terrorists. For a short time they began to force relatives of bombers to leave the areas assigned to Palestinian rule; however, the Israeli courts determined that was illegal and stopped the practice. While seemingly uncompassionate, the

Israelis have been limited in their actions by American influence. They have confined their use of assassinations to specific individuals. However, on many occasions bystanders have been killed during the attacks.

As harsh as this measure of killing terrorists' families sounds, it is based on experience. Many Americans, including CIA station chief William Buckley, were kidnapped in Beirut during the 1970s and 1980s. Only one was rescued alive. He is a friend of mine and had been working for the same agency. Very lucky Andre noted that "the calvary arrived at the last minute." However, the Soviets lost only three diplomats to Hezbollah terrorists. Without expressing outrage, the Soviets turned loose KGB trained-assassins, and dead Shiites were found brutally murdered. The pieces that were sent back told a story. The Soviet response was so severe that their personnel were never again abducted.[7]

Israel has used assassination as a tool with modest success. Their system requires that a death warrant be signed for a specific individual. Kidon teams, as the assassins are known, are then dispatched to eliminate their targets. Methods vary considerably. Many have just been shot wherever they have been found. At close quarters silenced .22-caliber pistols may be used. Other operations look more like military ambushes. For some, more ingenious means have been employed. At least a couple have been victims of telephones rigged with plastic explosive. Once the subject answers the phone and is positively identified, the set is detonated, taking off the terrorist's head.[8] Examples only, the manner in which assassination is accomplished is limited only by the imagination of the attacker. Ultimately, targeting for elimination specific people will save lives on all sides.

The conundrum we face is one of personal versus collective responsibility. Valuing individual rights so highly, we in the West tend to view acts in terms of the decisions made by each perpetrator. The notion of group responsibility is not new and has led to many wars. It was Osama bin Laden who issued a *fatwa* stating that all Americans, military and civilian, were legitimate targets. Greater extension of the concept is found in documents suggesting that payment of taxes to the U.S. government justifies a death sentence. The difference between my proposal and theirs is simply: There is an easy way out. If they don't commit terrorist acts, their groups, clans, tribes, and families have nothing to be concerned about.

There is a wrongheaded notion that all people fear being caught and brought before a court of law. There public exposure of acts and responsibility will be bared. But even if people are convicted at The Hague, penalties rarely fit the crime and the trials themselves make mockery of the suffering of victims. Long mute in the sanctum of death, they cannot bear witness to the unspeakable horrors that befell them. Rather, it must be the dispassionate testimony of forensic anthropologists and crime scene investigators who relay the technical aspects of their demise. Even mass murderers may serve only limited sentences while their victims remain dead forever.

There is a qualitative difference between indiscriminant terrorism that condones attacks on innocent civilians in public places and targeting supporters of suicide bombers. It is the terrorists themselves, and the leaders of terrorist organizations, who have prescribed the boundaries of responsibility. They target noncombatants, while their terms for participation in suicidal activities include continued support to their families. In so doing they have defined their relatives as part of the support mechanism of their illegal activities. It is therefore up to the terrorists whether they place their relatives at risk. The choice for the terrorists is simple. Stop the bombings and everyone lives. Continue those activities and the entire supporting element is eliminated.

There are inherent pitfalls in my proposal, ones that cannot be glossed over. To quote my friend Harry Rosen, "The danger in fighting dragons is that you might become one." Terrible decisions must be made and executed. We will have to send our best people to do the worst job imaginable. In the process, they will suffer greatly—and it is a price that will be extracted as long as they live. This is a function that should not be turned over to psychopaths, though that may seem the simple way out. The controls must be absolute, as such operations could easily get out of hand. It is a question of ultimate power. Who decides and under what conditions? As with nuclear deterrence, there must be clearly defined stopping points so that the adversary understands unambiguously your intent.

In the long run, a policy of genealogy-based extermination may save more lives of our adversaries and their supporters in countries of the Middle East. Following the attacks of 9/11 many people in the United States gladly would have supported a nuclear strike against the aggressors. Of course wiser minds prevailed, and it would have been difficult to identify an appropriate target for such a weapon. However, allowed sufficient time and support, terrorists will eventually conduct one or more attacks against U.S. interests that inflict enough pain and suffering that use of a nuclear system will be demanded. If the terrorists are not stopped before they reach our breaking point, it will become necessary for us to eliminate a country so that we regain respect.

Fear may not be a desirable choice for demanding appropriate deference and consideration, but it is a time-honored and effective one. If we fail to control our international relations and the situation devolves into a critical mass, we will have reached *Future War: The Event Horizon.*

PART IV

PLAN B: THE EVENT HORIZON

UNFORTUNATELY, WE LACK THE necessary political resolve and intestinal fortitude to do what is needed. Instead, we will go through several cycles of punctuated gains and losses. Our underlying philosophy in human rights and a belief that all people want the same things will cause us to take many obligatory actions. In the end they will likely prove futile. At some yet to be determined point in time, sufficient pain will be wreaked upon the United States. When that happens the battle lines will be redrawn and massively retaliatory response will be demanded. Should that occur, we will have devolved into the next global war—and it will be ignited along a religious fault.

Key to the future is what happens in the Islamic world. Many of the actions we take against those whom we define as terrorists will serve to further unify opposing forces. At present there is great instability in many countries. There are burgeoning populations, economic disparity, and increasing influence of fundamentalist thinking in the Middle East. Moderate leaders will have great difficulty in maintaining even a semblance of cooperative relationships with the West. The continued dominance of the House of Saud cannot be guaranteed. President Musharraf of Pakistan faces considerable internal strife for taking modest antiterrorist positions. His attempts at balancing Islamic interests while appearing sufficiently pro-West to win our concessions are doomed to collapse. Only when that happens is in question. President Mubarak of Egypt is seen unfavorably by many of his citizens. He has maintained his power by ruling through suppression of any opposition and receives a great deal of financial aid from the United States. Sentiment against him is rising as he attempts to align his son for succession. Indonesia, the fourth most populous country in the world and the one having the largest number of Muslims, is in a state of near-constant strife.

Provided Morocco remains fairly stable, the fate of Turkey may be pivotal. In the last century this ancient civilization transformed itself by accommodating the ways and technology of the West. However, there are now competing forces, including those that desire to adopt the more restrictive culture of their neighbors. The November 2002 elections swung power to the Islamic-rooted Justice and Development Party. Tension exists between them and the pro-Western military. The future of this strategic and culturally conjoined country cannot yet be determined.

Some observers believe that moderates have gained influence in Iran. At the same time there are contraindications including the discovery of al Qaeda camps there. Reports indicate a mutual support agreement between Iran and their traditional rival, Iraq. That does not bode well for actions that will surely take place in that region.

While geopolitical optimists postulate that Israel and its neighbors can coexist peacefully, I portend otherwise. Decades of open animosity fluctuate with constant insidious feuding interspersed by all-out wars. History does not lend credence to any theory that complex agreements involving ethnic islands beset by meandering borders will work. Without the necessary mutual trust and respect, extended peace seems unlikely. Omnipresent is the potential for conflagration with eschatological implications.

There are many moderate Muslims in the world. When asked about the nature of Islam they invoke the words of the Prophet Muhammad and point to the teachings of the Quran that speak of peace, mercy, and forgiveness. They also proclaim their reverence for the chain of prophets that includes those of Judeo-Christian beliefs with common ancestry traced to Abraham, Noah, and even Adam. Moderate Muslims note that Jesus is venerated as one of miraculous birth. According to their writings, they profess tolerance of other faiths as long as they are just.[1] Jihad, a word that evokes fear in many Westerners, is nothing more than "any struggle in day-to-day life to please God." In addition, "control of self from wrongdoings is also a great jihad."[2] It is clerics with such moderate opinions who appear with President Bush and other civic leaders to condemn terrorism and ask for tolerance so they can practice their religion in peace.

The problem is that this idyllic portrait of thoughtful and tranquil inter-religious relations does not reflect the real world. Around the planet there are many conflicts. The predominant ones are conflicts between Islamic fundamentalists and Jews, Islamic fundamentalists and Christians, Islamic fundamentalists and Hindus, and Islamic fundamentalists and everyone who doesn't agree with them—even other Muslims. The analytically minded reader can discern the critical pattern that emerges. To be sure, there are other conflicts in the world. But none, with the possible exception of an expansionist China, would involve global war.

There is an axis to be watched. Unshackled from the old East–West bipolar division we must now be concerned about a North–South rift. It runs from Morocco to the Philippines. Unlike the days of the Iron Curtain with its firm boundaries, these lines have been severely breached. Extensive immigration has allowed many Muslims to enter European countries. It is estimated that France alone has a Muslim population of over 5 million. There are about two and a half million Muslims in Germany. The United Kingdom has more than its share of radical Islamic fundamentalists. In fact, while most of the Western world held solemn remembrances on the anniversary of 9/11, celebrations were held in the streets of London to praise the suicide bombers. Thanks to liberal laws protecting freedom of speech, terrorist material banned in Cairo or Riyadh can be found in local UK bookstores. The permeation by people of Islamic faith in Europe will inhibit many political and military options in the future.

The Western Hemisphere is not isolated from the incursion. The triborder area of southern Brazil, northern Argentina, and Paraguay has a substantial Islamic population. The residents have established legitimate businesses that generate income. Money is then sent to support terrorists in the Middle East.[3] Highly placed sources indicate that the area also serves as a rest and recuperation venue for those wounded in fighting in other areas. Recently there has been a migration of terrorist supporters to other troubled areas of South America, including the border areas between northern Brazil and Colombia as well as in Bolivia.

Both the United States and Canada have increasing Muslim groups. The Islamic religion is growing rapidly in the United States, and estimates of the participants vary significantly. A recent Christian survey estimated there were about 1.6 million Muslims in the United States,[4] but Islamic publications all quote that number to be in excess of 6 million.[5] Deep in denial of realpolitik, the American clerics claim there is "no concept of Fundamentalism in Islam" and "there can be neither fanatic nor an extremist."[6]

In 1993 Samuel P. Huntington wrote the prophetic article "*The Clash of Civilizations?*" He opined that the dominating source of future conflict would be culture. The differing views on the relationships of God and man, individuals and groups, and citizens and states are a few of the essential points mentioned. Also noted were the increasing interactions that bring about awareness of differences between civilizations.[7] I am hardly the first to suggest that we have entered into conflict based on religion. Shortly after

9/11 Andrew Sullivan wrote a chilling article for the *New York Times* stating the war had a religious basis. However, he attempted to ameliorate the sonorous headline by claiming the conflict was not between Islam and other religions.[8] Similarly Pope John Paul II addressed terrorism and mentioned a "clash of civilizations that at times seems inevitable."[9] It is: not with the philosophically pure Islam but the real-world Islam.

An important concept that needs correcting is the notion that everybody wants the same things. Americans have the misguided idea that everybody wants more high technology and a comfortable life. That usually translates to having more things. The success of blue jeans and rock music serves as an example. There are serious lessons to be learned from Thomas L. Friedman's *The Lexus and the Olive Tree*[10] and Robert D. Kaplan's *Warrior Politics*[11] that suggest we do not understand the conflict in which we are engaged.

In discussing this topic with Tom Clancy, he said that following 9/11 he contacted Secretary of State Colin Powell, now a cousin by marriage, and told him we needed to make the Arabs rich. Clancy's premise was that if physically satisfied, the moderates of Islamic countries would control those potential terrorists. Their influence would be less likely to spawn more of that kind. I opined that there are very substantial differences in the belief systems and increasing physical comfort alone would not alter the course of history. Which one of us is right has yet to be determined. However, understanding these issues is critical to establishing a basis from which this war can be won.

The notion of internally competing beliefs, articulated in Appendix A of *Future War,* will severely test our patience, trust, confidence, and most particularly the laws of our countries.[12] It will be at this juncture that the trade-offs between security and rights will clash inextricably. Until now there has been an assumption that citizenship established the dominant allegiance of soldiers. Religious beliefs supported participating in conflict or seeking conscientious objector status. But then, we have never been asked to define an adversary based on religion.

The first step will be to redefine the conflict. In the appendix is an Op-Ed piece that I wrote and was published in the *Las Vegas Sun* on October 18, 2001. The concept remains valid today. Defining the issue is an integral part of winning the conflict. The War on Terror was defined too narrowly to facilitate victory. We made a valiant attempt to define the war as related only to terrorists. We excluded those of the Islamic faith who did not support terrorists' acts. However, the persuasive powers of fundamentalist Islam will most likely prevail, at least with the 1 billion Muslins in the world. These fundamentalist leaders choose to define themselves as the opposition. If they are successful, our control of the initiative is lost.

The second step is to reconcile resources available for the fight. We simply cannot afford to buy off every potential adversary. Worth repeating is that Afghanistan has no reason to exist as a nation-state. Yet we support the effort to make it one. They lack a common language, common heritage, and common purpose. Socially their tribes have rarely risen above feudal squabbling. The American taxpayer is asked to fund the nation-building efforts that are doomed to failure. Economically and militarily we do not have sufficient resources to fight all of these battles, maintain stability in a multitude of places, and float multinational alliances that hold only as long as we pay the bills.

The third step is to prepare to go it alone. For the reasons stated earlier, our traditional allies will have difficulty supporting the actions that must be taken to eliminate tyrants with weapons of mass destruction. Europe will pull back to deal with internal problems caused by pandemic immigration. Depending on resolution of their economic issues, it may well be Russia that we come to rely upon on that continent. As the Israelis have learned, when things get really tough you have to be self-sufficient.

The foundations supporting the *War on Terror* already show signs of crumbling. There is a distinct shift toward high-tech warfare at the expense of ground troops. Soldiers in Afghanistan report that they buy equipment from Wal-Mart out of their pockets because it is better than issued items.[13] Importantly, there are not enough troops in the military to accomplish all of the missions required of them. We have lost sight of strategic thinking that must address enduring solutions. There is no doubt we can blow things up and destroy any traditional armed adversary. However, maintaining peace requires more than destruction of enemy leadership and threats of using additional force. Frequently that means long-term commitment of ground forces and sustained economic assistance. The extent of these requirements should be determined prior to use of force.

Our notion of propriety needs to be reconsidered if we expect to win future wars. Articles on combat in Afghanistan complain about bureaucrats and lawyers being politically correct and placing great restrictions on the actions of Special Operations forces in that theater. They claim that "if you put in a conop (concept of operations), if it said 'raid,' 'ambush,' 'kill,' 'sniper,' anything like that, the conop would be disapproved based on the vocabulary used."[14] These actions are being taken due to concern over collateral casualties. The article continued: "The fear of getting prosecuted for anything is real. There's a paranoia, there are so many lawyers." Such political correctness will have unacceptable consequences. The future is pain-mediated. As the pain threshold rises, we will be forced to choose between politically correctness and national survival.

This part of the book provides potential solutions for the go-it-alone eventuality. They are not the only solutions, but they are a start. An asymmetric threat deserves an asymmetric response. We can alter the equation by addressing alternative energy production as our primary goal. This will also minimize the philosophical clash of Islam as direct interaction decreases. The focus of the proposal is twofold: become free of petrochemical dependency for energy and dominate space. That does not mean *assuring access to space* but rather taking the high ground and dominating it—absolutely.

When everything goes to hell we revert to contingency plans. It is not yet time to implement it, but I'd like to think somebody is developing Plan B. However, if we do need to execute it, we are looking at the event horizon in warfare.

CHAPTER THIRTEEN

RETHINKING SPACE MISSIONS

"AMERICA'S FUTURE SECURITY AND prosperity depend on our constant supremacy in space." Those words by Senator Bob Smith of New Hampshire are even more important today than when he published them in 1999.[1] Powerful forces are at play, ones that will permanently alter U.S. military involvement in space missions. At least two significant factors will drive that change. The first is geopolitical and involves the *War on Terror*. We simply cannot rely on expensive, transitory, and unreliable alliances in order to project and apply power. Operating from space frees us of those constraints. Second, based purely on civilian economics and entrepreneurship there is likely to be a substantial manned space presence in time lines well under those currently anticipated. Nearly all of the long-range planning (out twenty years) conducted by the U.S. Space Command assumes that with the exception of the International Space Station (ISS) space utilization will be exclusively unmanned satellites.[2]

For several decades there has been debate about the efficacy of weaponizing space. Starting in the 1950s the superpowers began competing for domination of space. There were two key issues: national prestige and intelligence collection. Against this backdrop President Eisenhower set national space policy and the United States became the proponent of freedom of space. It was his view that it was in the best interests of the United States that weapons not be deployed in space.[3] However, that was in a bipolar world in which deterrence and containment, a concept outlined by George Kennan and Paul Nitze and published in the secret document NSC 68, dominated all foreign policy decisions. For the following decades, great care was given to actions that might destabilize the delicate balance of power. It is highly unlikely that those policy makers could have envisioned either the proliferation of space capabilities currently available or the dramatic shifts in global power structures.

WHY WE NEED SPACE

All U.S. military forces benefit from the available satellite-based capabilities. Aerospace integration is key to both current and future warfighting capabilities.[4] The U.S. Space Command postulates the need to dominate space.[5] However, we are being challenged. As Senator Smith and others rightfully have pointed out, continued U.S. dominance of space is not assured without "sustained and substantial commitment of resources."[6]

The terrorists' attacks of September 11, 2001, and subsequent demarcation of the *War on Terror* by Pres. George W. Bush have provided cause to rethink weaponization of space.[7] Operation Enduring Freedom is a textbook case for the problems associated with conducting military operations in places that are difficult to reach. Despite the USAF motto of "*Global Reach, Global Power*," striking targets rapidly in Afghanistan was untenable. As the country is landlocked, the United States negotiated various overflight and basing rights before any missions commenced. That took nearly a month. Even then, the B-2 bomber strikes required extraordinary efforts on the parts of multiple crews that flew more than seventy hours round-trip from Whiteman Air Force Base via Diego Garcia.[8] Before other strikes could begin, significant logistical efforts were conducted to locate aircraft and support matériel in the region. Except for long-range bombers, most air strikes were U.S. Navy carrier–based aircraft.

In every case, the development of the alliance was at significant financial and political cost. Pakistan had sanctions lifted and received a billion-dollar aid package.[9] Deals were made with Russia, Uzbekistan, Tajikistan, and other countries for basing and overflight rights. For the sake of expediency, inside Afghanistan the United States became embroiled with ethnic factions with unsavory backgrounds and competing postconflict agendas. Even countries such as India and Sudan, though not directly involved in the conflict, received special consideration. Perhaps most problematic is that some of the complex agreements curried favor with governments that were inherently unstable due to internal strife and could lose power with short notice. With such transitions, critical agreements could be instantly negated.

One has but to look at the accelerating rate and dramatic impact of the geopolitical changes of the last fifteen years to understand the uncertainties of relying on terrestrial alliances. Extant treaties notwithstanding, strategically, space-based weapons and absolute domination of that realm are in the nation's best interest.

THE PROBLEMS OF ASSUMPTIONS

There are serious problems with the assumptions that the space domain of military interest will remain unmanned. Conventional wisdom, driven by a myopic NASA view, suggests that the future of commercial space is limited to satellites. It is perceived the ISS as the only manned platform and established policies that tend to support that position. In the NASA view, privatization of space infers that companies enter into contractual agreements with NASA-funded commercial space centers to use the facilities of the ISS. Given very limited availability of research time and space on the ISS, competition for access will be extreme.

In general it is assumed that the costs associated with space launches will limit the participants to nations with advanced technology and large capital resources. As Maj. Gen. Barry and Colonel Herriges rightfully point out there is a "cost-to-orbit challenge." The rule of thumb mentioned—$10,000 per pound—has held for the past twenty years.[10] Should that remain a valid cost estimate, then space enterprises will remain relatively restricted. This view has impacted our planning for protecting our assets in space. As Lt. Gen. Eugene Tattini stated, war in space will be "all done remotely," with directed energy and sensors.[11] However, both countries and private enterprises are conducting extensive research and development that may alter the cost-to-orbit equation. A breakthrough in any one of several areas could bring a resurgence of entrepreneurship with startling consequences for the space domain.

Foreign competition for space launches is already making inroads. Because of the availability of lower-cost options, American communications companies routinely have payloads sent into orbit from China, Russia, France, and other countries. Because of the internal economic considerations, their cost of doing business, especially in personnel payroll and benefits, can be far below those of the United States. Further, China has announced its intention to establish a manned presence in space as soon as possible and set lunar landings on the agenda.[12]

American businesses could also alter this market. To date, the NASA-dominated space industry has acquired a reputation for poor management and willingness to pay extremely high prices for the products they need. Many NASA projects have been severely over budget, which has drawn the attention and ire of Congress. As an example, it was learned in the summer of 2001 that the ISS faced a $4.5 billion overrun. To adjust these costs it was recommended that science experiments, the primary purpose for the

ISS, be curtailed.[13] Because of an absolute monopoly, NASA failed to employ business practices that are standard in the private sector.

It is likely that improved business practices could be a major factor in reducing the total cost-to-orbit challenge. The savings would be found across the entire spectrum of the space enterprise. There will be reductions in the design, fabrication, and deployment expenses. With good cost-conscious business practices, expenditures could be brought down significantly. Such a drop in overall costs would make space-based business attractive to those who cannot currently afford the high price associated with getting their products into orbit.

Another likely possibility is an advance in space platform design that would reduce the number of total launches necessary to place it in orbit. NASA has explored the possibility of employing inflatable modules called TransHab.[14] Though they terminated work on that technology, others in the private sector did not.[15] Currently research is well under way in the development of prototypes of inflatable space stations.[16] Even though constrained by existing payload capacity, it is estimated that inflatable modules could provide 2.7 times the cubic space of a typical rigid module on the ISS. If the size of launch payload capacity is enlarged, the advantages of employing inflatable modules increase substantially. More important, the number of total launches required to orbit a complete inflatable space platform is estimated to be as few as two, rather than the seven needed for a comparable but much smaller NASA-like space platform. The baseline cost per inflatable module from the private sector is estimated to be $50 million, compared with the current ISS modules at $750 million to $1 billion.[17]

While cost to orbit is an important factor, the combination of inflatable platform technologies and reduction in number of launches provides another important metric. Instead of limiting consideration to the expense of launch, the cost in cubic meters (m^3) of habitable space can also be compared. When the ISS base costs plus seven launches yielding 110 m^3 is compared with inflatable base costs and two launches providing 300 m^3, inflatables stations have a 30:1 advantage in cost for habitable space. That factor should have significant impact on interest in space for commercial purposes.

There is a likely combination of better business practices, improved design, and innovative technology, thus reducing total lift requirements. These factors would dramatically alter availability and expand the number of organizations that can enter into space enterprises. In fact, in October 2001 the first private U.S. company applied to the Federal Aviation Administration for a license to build and launch a three-module inflatable space station.[18] The technologies involved in the development of inflatable modules are well known, and therefore the risk factors are low and manageable.

By substantially reducing the total cost-to-orbit of a space platform, access will be afforded to many new countries, organizations, and possible individuals with the financial means to participate. Comparable to the introduction of cellular phones to telecommunications, development of inflatable modules allows access to space without gigantic investment in an aerospace infrastructure. Components could be bought off-the-shelf and launches contracted from existing facilities.

Both the public and military sectors have been exploring development of reusable launch vehicles (RLVs). The stated goal of the NASA X-33 program was to reduce launch costs to approximately a thousand dollars per pound.[19] A detailed discussion of reusable launch vehicles was written by Lt. Col. John E. Ward, Jr., and is available in *Occasional Paper 12* of the Center for Strategic Technology, Air War College. There are significant technical challenges in operability and reliability, and the developmental costs would be quite high. It is noteworthy that Lieutenant Colonel Ward sided against the U.S. Air Force involvement in research and development of RLVs, but predominantly for political reasons. In his view, military use might be a destabilizing factor and start a race to weaponize space.[20]

The X-33 and X-34 programs were terminated for unacceptable technical risks and budget problems in March 2001.[21] These were known to be embryonic attempts; thus high risks were anticipated. However, it is likely that research into RLVs will continue and the successful development in reduction of cost to orbit clearly would change the market and bring many more civilians into space.

Yet another possibility is a breakthrough in propulsion systems. Though less likely than business improvement, the U.S. Air Force, plus many corporate and private researchers, are diligently pursuing promising technologies. In addition, NASA has formed a multiagency *Breakthrough Propulsion Research Group* that has been funding projects for several years.[22] In the event that one of these is successful, then cost per pound could drop significantly.

WE MUST CHANGE ASSUMPTIONS AND CONSIDER NEW OPTIONS

Current assumptions perilously ignore even the potential for a leap in technology that would lead to greatly expanded manned presence in space. It is believed that access to a safe, affordable microgravity environment will generate commercialization in space on a scale far beyond what is planned for the ISS. Since we are still in an embryonic stage of space flight it is hard to envision all of the opportunities that will be presented to us. Certainly the Wright brothers could not have predicted the multifaceted applications of the modern aerospace industry.

Initially there will be requirements for basic research into areas that cannot be studied in a terrestrial setting. Topics to be investigated will include fluid physics, combustion, nanotechnology, human physiology, biology, thin films, light-scattering diagnostics, electrohydrodynamics, and thermocapillary convections, to name a few.

In addition to basic science it will be the economic benefits of unique manufacturing capabilities that will draw private industry into space. Already pharmaceutical companies are busy preparing for experiments that can only be done in space. There are high expectations for the manufacturing of protein crystals needed by that industry. In a recent study the National Academy of Sciences noted that "despite the lack of impact of microgravity research on structural biology up to now, there is reason to believe that the potential exists for crystallization in the microgravity environment to contribute to future advances in structural determination."[23] The study focused on ISS research, but it can be inferred that the amount of effort required would greatly exceed ISS capacity.

Some concern has been raised about whether delicate manufacturing processes can withstand the additional vibration caused by human life-support systems. In fact, sensitive operations may be conducted in separate platforms that are either tethered or stationed in very close proximity. This would allow rapid inspection by humans with harvested materials protected and shipped by bus before transportation to Earth.

Many other manufacturing opportunities await availability of space platforms. Some of these include microencapsulation for cell-based therapies for treating kidney or liver failure, production of zeolites for use in the chemical industry, plus aerogels for insulation and nonlinear optics.[24] These are but a few of the technologies that will draw private industries into space as soon as economic alternatives are available.

In addition to manufacturing, proliferation of manned space platforms will follow for other economic reasons. They would make a port of embarkation for more extensive space exploration including expeditions to the Moon, Mars, and beyond.[25] Space tourism, which began in 2001 with Dennis Tito's trip to the ISS, will expand. High-cost vacations are no longer an anomaly, and demand will exceed supply for some time.[26] Mining of near-

Earth asteroids may be possible. Consideration has been given for development of solar power stations that could send energy back to Earth.

There are national benefits for establishing a strong space presence. These include enhancing U.S. industry competitiveness, creating spin-off technologies for nonspace industries, and increasing national prestige.[27] Former Vice Chief of Staff U.S. Air Force General S. Moorman, Jr., noted that even in the satellite-dominated market there was an explosion in commercial space and "the rate of change will become even more dramatic."[28] However, the driving factor in commercialization of space will be economics. If significant advances or technical breakthroughs occur in one or more of the areas mentioned, manned space presence will burgeon forth very quickly.

NEW MISSIONS

A significant manned presence in space will offer opportunities for military use of space and demand new missions be established. It clearly prescribes an environment very different from the one assumed during the years of debate concerning the sanctity of space that have transpired. Some have argued that space is a sanctuary and should remain free of weapons.[29] Others have countered that weaponizing space is inevitable.[30] None, however, have considered the implication of military operations in space with a significant number of high-value assets in the form of manned space stations.

While domination of space will remain the fundamental mission, how that is performed will undergo a radical change. It is not likely that human temperament will change just because of the new venue. Therefore, an organization will be required for policing the space domain. There will be a need for monitoring and control of location or orbital patterns of the platforms, plus the requirement for physical protection. As certain activities will be banned, it is likely there will be a need to monitor the commercial endeavors of the private habitats to ensure compliance with international treaties and trade agreements.

Rescue has traditionally been a mission accepted by governments. As space is a high-risk environment, there is a strong probability that accidents will occur and loss of life must be anticipated. Whenever possible, rescue attempts will probably be made, and the military is the most likely agency to carry them out. The debate about the structure of such units has been initiated. Whether it will remain a U.S. Air Force function or a Space Guard will be formed is not clear. But some military organization will be assigned the mission.[31]

In addition to policelike actions, the military will choose to use space for one or more of the following purposes: a manned presence will allow military research that parallels the basic science advances in the civilian sector. Military laboratories have a long and distinguished history in selected scientific fields. They will continue that legacy in space.

The existence of relatively large space platforms will also permit retrieval and repair of smaller unmanned satellites, thus greatly extending their usefulness. In addition, as technical advances are now evolving more rapidly, the capture and upgrade capabilities will increase the use of preplanned produced improvement or block modification for surveillance or communications satellites. When needed, these manned platforms could be used for sophisticated intelligence gathering, analysis, and dissemination. The volume available would provide for capabilities beyond those on one or more unmanned satellites.

Parked in geosynchronous orbits or at Lagrange Points, manned stations could provide very secure command and control facilities in the event of major war or devastating terrestrial calamities. These events might include major geophysical changes, large meteor impacts, or the hypothesized superstorms. At that distance they could be protected from any Earth-based kinetic or directed energy antisatellite (ASAT) weapons or physical side effects from natural disasters. As a space-based version of *Looking Glass*, they should be assured control of space, airborne, and most naval assets in time of conflict even if traditional ground downlinks have been negated. Locating that base on the backside of the Moon with relay stations for communications would ensure a very secure site.

More contentious would be emplacement of manned and remotely controlled weapons platforms. They could deploy or threaten to deploy weapons of mass destruction to deter aggression or to prevent strategic defeat. Both directed-energy and kinetic-impact weapons could be employed with pinpoint accuracy, promoting controllable levels of damage designed to meet specified threshold criteria established by military and political planners. Most important, they would be able to respond almost anywhere on Earth within minutes, not the weeks we currently require. Retired Air Force Chief of Staff General Ronald R. Fogleman has already predicted, "that by 2020 directed energy will be the centerpiece of the U.S. military arsenal."[32] Some of that will probably be space-based. Such platforms could either designate targets or provide coordinates for airborne precision-guided weapons.

Planetary or asteroid defense is a mission that has been considered by the military.[33] In fact, there is a joint Naval Postgraduate School and Air Force Institute of Technology Web site devoted to the topic.[34] Operating from a space-based platform could enhance the capability to detect near-Earth space-borne threats.[35] Edward Teller stated that the failure to recognize the threat was a scandal and there was danger of a strike occurring with no warning.[36] At the same conference it was noted that the most cost-effective method to interdict an asteroid would be a nuclear explosive.[37] Greg Canavan, speaking before the Congressional Hearings on Near-Earth Objects and Planetary Defense, indicated that we have not yet addressed the problem of long-period comets whose orbits intersect Earth's and this is a significant problem.[38] The seriousness of this problem was noted when an asteroid large enough to wipe out a major country passed close to Earth on January 8, 2002. Named 2001YB5, it had not been detected by the Near Earth Asteroid Tracking (NEAT) survey telescope until December 2001.[39] Still worse is that new, previously undiscovered asteroids keep popping up. Sometimes they remain unknown until after they have whizzed by.

From multifunctional space platforms the capability of early detection would be increased. If it was necessary to attempt to interdict or deflect that threat, initiating that attack would be accomplished more easily. Even honoring treaties to prevent keeping nuclear weapons in space, fueled interceptors could be maintained on these bases and warheads delivered upon confirmation of extraterrestrial threats.

NEED TO REEXAMINE MANNED SPACE REQUIREMENTS

No study to date has seriously considered the requirements for the U.S. Air Force to operate in a manned space environment. The simple assumption that the economic factors would be too great to permit a substantial human presence in that domain was the predominant thought. However, as expressed in this chapter, several current research and business initiatives bring that underlying assumption into question.

It is likely that civilian enterprise, not the U.S. government bureaucracy, will alter the path of commercialization of space. These entrepreneurs understand the admonition from Col. John A. Warden, USAF (Ret.), when he said, "You won't win by following the rules of yesterday."[40] The aforementioned combination of better business models, innovative application of existing technologies, and economic incentives will greatly accelerate the time frame in which manned space platforms will proliferate.

Therefore, it is time for the U.S. Air Force to critically reexamine their operational requirements should the capability for manned presence mature in years, not decades. The planning process must be comprehensive, including strategic, operational, and tactical considerations. Extensive examination must be given to the skills necessary for sustained deployments in space. Further, the United States has been slow to delve deeply into topics such as personality, interpersonal relationships, and group dynamics as they relate to space missions. As Al Harrison states in *Spacefaring: The Human Dimension*, "Changes in crew size, crew composition, mission duration and technology prompt us to rethink all aspects of spaceflight."[41] Given the new missions that will evolve, existing astronaut selection and training will probably be insufficient to meet those needs.

Concurrently there is a need to begin in-depth exploration of the various possible configurations that the U.S. presence in space might take. It is assumed that the era in which space was considered a sanctuary has ended.[42] While we are willing to share the domain with other peaceful space-faring nations and enterprises, national security interests should be preeminent in our selection of space policy and organization.

The actions suggested in this chapter will be met with scathing criticism in some circles and especially from arms control proponents. These actions represent a significant departure from the established *National Space Policy* in that they place heavier emphasis on the national security aspects than international cooperation. Written in the Cold War era, that policy was designed to encourage the use of space while attesting to mutually beneficial applications for defensive purposes. Carefully maintaining the existing balance of power was a dominant consideration.[43] The problems associated with asymmetric conflict were not adequately evaluated.

However, the recommendations in this chapter are commensurate with the findings of *The Space Commission* headed by Donald Rumsfeld before he again assumed the position of Secretary of Defense. The commission concluded that it was in the U.S. national interest to "develop and deploy the means to deter and defend against hostile acts directed at U.S. space assets and again the uses of space hostile to U.S. interests." Among the specific military capabilities they recommended were:

- Global command, control, and communications in space
- Defense in space
- Homeland defense
- ***Power projection in, from, and through space*** (emphasis added)[44]

While we as a country chose to promote peaceful uses of space, there is an agreed-upon need "to ensure the United States remains the world's leading space-faring nation."[45] We must not be dissuaded by the anticipated criticism of U.S. dominance in space. Our national security depends on it. Worth noting are the words of former Secretary of Defense Caspar Weinberger when appearing on *Larry King Live*. He stated, "The thing that people around the world are always talking about is a too strong America and all that. What they really fear is a weak America."[46]

SUMMARY

The past decade has seen major changes in the geopolitical landscape. While alliances may work in some situations, they are difficult to establish and maintain, expensive, time-consuming, and potentially unreliable. Threats in the future may not allow the time and effort necessary to build coalitions in order to attack elusive targets.

Concurrently other countries are making advances in space technology and our dominance cannot be assured without extensive commitment. There is reason to believe that a substantial manned space presence may be much closer than anticipated in Space Command's long-range planning guidance. This will be driven by economic considerations in the private sector. If U.S. companies don't establish private space platforms, other countries will. It will not take a big leap in technology, just ingenuity and solid business practices.

Two issues will dominate the role in space. One will be whether exploration and other functions will be conducted remotely or with manned platforms. In reality it will be a yet to be determined mixture. At this moment in history we have the ability to maintain dominance of space. That lead will not last very long. Though politically unpopular, thorough examination of the options is necessary. The second related issue will be the economic impact provided a new cost paradigm. It is likely that true civilianization of space will alter who leads the expansion of space exploration and exploitation.

Increased manned space exploration has been a very controversial issue. However, on April 18, 2003, legendary aircraft designer Burt Rutan shocked the space industry by unveiling a totally civilian enterprise that had been developed in a shroud of secrecy. With lunar astronaut Buzz Aldrin present, Rutan displayed an *entire manned space program*. Complete with SpaceShipOne, a launch vehicle called the White Knight, and the ground support stations, this program was beyond the wildest imagination of space experts. More importantly, he had done it on a budget three orders of magnitude lower than NASA. Rutan's goal is to make space accessible to everyone.

Therefore, it is time for the Air Force to begin serious consideration of their roles and missions in the space domain that is inhabited. There is also a need to have extensive discussions concerning the appropriate role of the U.S. military in a venue shared by other nations. Given our current ability to control the next phase of development of space, what will be in our nation's best interests? The questions are complex, but now is the time to study them seriously.[47]

CHAPTER FOURTEEN

SIX SIGMA SOLUTIONS

IN THE ASYMMETRIC CONFLICTS in which we are currently engaged, we must be looking for the big leaps: technologies or concepts that push the envelope and change the equation. Popular in business practice is the notion of six sigma manufacturing.[1] The notion entails establishing exceptionally high performance standards for situations in which nothing less will do. For instance, a 99 percent success rate in flights is not good enough for the airlines industry. A 1 percent error rate would mean as many as 170 flights per day with major problems or crashes. That 1 percent failure rate would be sufficient to keep people off airplanes.

A statistical term, *six sigma defects per unit* indicates a success rate greater than 99.99966 percent—in other words, virtually defect-free, just as we want the airplanes to be. For our purposes, six sigma solutions are long shots—high-risk technically but with an extremely high payoff. They provide innovative concepts that can change the very nature of the way we prosecute conflict. These will be essential in asymmetric warfare that is being conceptualized and waged by our adversaries.

In his "Navy After Next" briefing to all U.S. naval flag officers, Vice Admiral Cebrowski noted several factors that will define conflicts.[2] First, the nations-at-arms concept is defunct. War is likely to be waged by non-state actors, with and without nation sponsorship. Second, there is a decoupling between the amount of destruction inflicted on an adversary and victory. Unlike the case in World War II, the focus is not on destroying the enemy's country to be declared the winner.[3] Also, weapons of mass destruction will be used against governments and societies.

With this change in focus of warfare, it should not be surprising that the six sigma solutions are not necessarily weapons systems. Rather, they are innovative technologies and human capabilities that have the power to alter the path of civilization.

ENERGY

As mentioned in "The War We Will Fight," energy is at the core of our foreign policy. Third World countries that have large reserves of petrochemicals fare far better than those that do not. The technologically developed countries currently are inextricably linked to those possessing large quantities of fossil fuels. Therefore, any breakthrough that provides the potential for freedom from the oil harness has strategic value.

There are many research projects on alternative energy sources ongoing. Sooner or later one or more are likely to provide the answer necessary to break the chain. Only then will we stop the chant *"No blood for oil."* Here are a few examples. Some are a bit more speculative than others, but all have sound science behind them. Further, advances in the more conventional efforts in renewable energy sources will help us reduce dependence on external sources and thereby play a significant role in our future national security.

FUSION POWER

An alternative form of nuclear power called fusion power promises to be free of radioactive waste, use a fuel that has an unlimited supply in seawater, and be safe even under conditions of extreme malfunction.[4] In order to understand fusion power, it is necessary to compare fossil, nuclear, and fusion power basics. All fossil fuels release energy by chemical reactions. The energy that can be realized from oil, natural gas, and gasoline is about 40 MJ/kg (coal is about half this value). Two types of nuclear reactions can produce useful energy: fission and fusion. Fission is the splitting of a heavy atomic nucleus such as uranium or plutonium into lighter elements whose sum of masses is less than the mass of the original nucleus. This mass deficient is converted to energy by the famous equation of Einstein that $E = mc^2$, where c is the speed of light. The energy release in the fission of the nucleus is about a million times larger than the energy release from chemical reactions. Consequently, a kilogram of uranium can produce a power level of 1,000 megawatts of electricity for a day or an energy release of 86 million MJ/kg of fuel. This is the reaction that powers today's nuclear plants. Because of this million times difference in energy release between chemical and nuclear reactions, a nuclear power plant can produce the same amount of power as a fossil plant but with much less fuel.

Fusion reactions occur in the light elements. As the name implies, two light elements fuse together to form a heavier element where the mass of the initial elements is greater than the heavier element product. By the same

Einstein equation, the conversion of mass into energy by fusion can be very large. A typical fusion reaction occurs when two hydrogen atoms fuse to form helium. However, for this reaction to occur the temperatures must be increased to nearly 100 million degrees. It is certain that one can produce energy from fusion reactions because it is the hydrogen fusion reaction that is responsible for the energy release in the Sun over its life span of 5 billion years and the almost instantaneous energy release in an explosion of a hydrogen bomb.

The idea of fusion power is to heat up the light element fuel to 100 million degrees and control the fusion energy release in such a way as to drive a thermal power plant. In any thermal power plant, fossil fuel, nuclear, or fusion, the heat released by the fuel is used to produce high-pressure steam and drive a turbine, which in turn drives an electrical generator. The fuel for proposed fusion reactors is a heavier form (isotope) of hydrogen called deuterium (or a third heavier isotope tritium). The reason for using deuterium is that it fuses or "burns" at a lower temperature than hydrogen fusion reactions. Deuterium is present in water in a D/H ratio of 1/6400, but it is easily extracted and the supply is virtually unlimited. The energy that can be extracted by fusion reactions is about 0.67 million MJ/kg of fuel.

Two very different methods currently are being considered to heat and control fusion energy reactions. One is magnetic confinement and the other is inertial confinement effected through the use of high-powered lasers, also called laser fusion. Scientists have been working on controlled fusion and confinement for over fifty years; strong programs in both areas increased dramatically in 1974. Yet the physics of fusion heating and containment has proved difficult and has produced many surprises along the way of the research. These discoveries have had the effect of requiring larger and more expensive machines to achieve a condition of breakeven, a condition where the energy generated by fusion is greater than the energy used to initiate and contain the fusion reaction.

The longest and largest efforts have been in magnetic confinement fusion. In this concept, a low-pressure gas of deuterium is heated to approximately 100 million degrees by a number of different methods. At these high temperatures, the electrons are torn apart from the atoms to form a gas of charge particles called plasma-electrons and deuterium ions. No elemental material can contain this plasma at these extreme temperatures. However, the charged particles can be trapped and contained by strong magnetic fields. Magnetic fields confine these high-energy-charged particles to various geometrical volumes formed by the magnetic field, a configuration sometimes referred to as a magnetic bottle. One such configuration is in the form of a torus that is called a Tokamak (named by the Russians who discovered this particular configuration). This has been the favored approach for magnetic confinement fusion for the last two decades.

The potential of fusion as an ideal energy source to produce electricity is so great that almost all of the developed nations of the world have major programs in fusion research, primarily in Tokamaks. In the United States, there is a major effort at Princeton University with the Tokamak Fusion Test Reactor (TFTR), at MIT, and at General Atomics in San Diego, California. A European consortium developed the Joint European Torus (JET) and announced major progress in 1991. The Japanese (JT-60) and Russians (T-10) have had their own programs in this area. As the size of these projects grew to the several-billion-dollar level for the next proof of concept machine, a global venture was formed called the International Thermonuclear Experimental Reactor (ITER). The major nations initially involved in this effort were the European nations, Japan, Russia, and the United States. However, decisions around 1990 were made to de-emphasize the U.S. magnetic fusion program and concentrate on the inertial confinement approach. The United States now plays only a minor role in the ITER program.

The inertial confinement approach to fusion, simply stated, is to create a miniature "star" in the laboratory. The generation, heating, and confinement of the deuterium plasma is provided by very high-power lasers that irradiate and compress a small pellet containing the fuel. The containment is inertial in the sense that the fusion energy would be released in a time short compared to the time for the highly compressed pellet to fly apart. This process would occur in less than a billionth of a second but could be repeated ten to a hundred times a second. During each laser pulse, a fusion reaction would occur that would generate neutrons that would be captured in the walls of a vessel and turned into heat. This heat would be used to generate steam to drive a conventional thermal power plant.

The challenges for such a concept are multiple. The fuel must be compressed to about twenty times liquid densities and heated by compression to 50 million degrees. By comparison, the conditions at the center of the Sun are 20 million degrees and about a factor of ten lower pressure than is required in the compressed fuel pellet. The lasers required to compress the fuel pellet are extremely large in scale and cost.

The major effort in inertial confinement fusion is in the United States, primarily at the Department of Energy National Laboratories and at the University of Rochester in New York. The largest effort is at the Lawrence Livermore National Laboratory in California. A laser has been built that generated an energy of 100 kilojoules in a pulse of a billionth of a second (NOVA). Currently under construction is the National Ignition Fusion Facility (NIFF), which will generate energy in excess of megajoules with multiple beams. However, the cost of the laser project has escalated to several billion dollars, so its future is uncertain at this time. In various experiments, compressions and temperatures have been achieved that are comparable to what is required for breakeven fusion.

Fusion power has great potential as a major source of energy for the growing needs of the world. It is non-polluting, has no issues with generation of carbon dioxide and global warming, and can use a fuel that is virtually unlimited in supply and is easily accessible. However, the scientific, technical, and engineering challenges are daunting. Major efforts are in progress in many countries of the world to achieve success in power generation through thermonuclear fusion. With such major efforts in progress and the best scientific talent bent on solving the problem, fusion power will one day become a reality. The question is when and at what economic cost of the power to the customer.

ZERO POINT ENERGY (ZPE)

To the non-physicist, empty space, say the vacuum of outer space, would seem to be truly empty. And with regard to most observation it is, with essentially no atoms or molecules to be found. As a result empty space can for all practical purposes be considered a void. And it has so been considered by physicists as well, at least until the advent of quantum theory.[5]

Surprisingly enough, it was not experiment but theory, quantum theory to be exact, that indicated that empty space is not truly empty but rather is an underlying sea of random fluctuating quantum processes, with particles arising and disappearing, awash in a sea of fluctuating electromagnetic fields. The predicted unexpectedly large intensity of these effects caused physicists initially to be concerned about whether the mathematics of quantum theory might be flawed. (Energies on the order of nuclear energy densities or greater were being predicted!) However, careful experimentation soon revealed that the predicted phenomena were indeed real, and their confirmation resulted in a Nobel Prize for one of the early experimentalists. A recent very accurate measurement of these effects resulted in a *New York Times* headline "Physicists Confirm Power of Nothing, Measuring Force of Quantum Foam," with a subtitle "Fluctuations in the Vacuum Are the Universal Pulse of Existence."

The vacuum fluctuation energy is typically referred to as zero point energy, or ZPE. The adjectival *zero-point* signifies that such energetic activity remains even at the zero point of temperature (absolute zero) after all thermal effects have frozen out.

Originally it was thought that ZPE concepts were of significance only for such esoteric concerns as small perturbations to atomic emission processes. It is now known, however, that the all-pervasive energetic ZPE fields play a central role in large-scale phenomena of interest to technologists as well, thus leading to the possibility of "vacuum engineering." One area of interest is the enhancement or inhibition of the emission of radiation from atoms and molecules, of importance to microwave and laser engineers. Another is the generation of short-range attractive forces between closely spaced materials, of significance in the design of so-called nanomachines. There is also the possibility of extracting useful energy from the vacuum fluctuations of "empty space," that is, "mining" ZPE for practical use. If this could be done, it would constitute a virtually ubiquitous energy supply, a veritable "Holy Grail" energy source. Then there is the potential for modifying inertial and gravitational forces, in which case "warp-drive" space propulsion could enter the realm of scientific possibility. As recently stated in a review of the field: "The vacuum is fast emerging as the central structure of modern physics."

ANTIGRAVITY

Antigravity is the ultimate envy of the propulsion and transportation industries.[6] Any form of propulsion that is based on antigravity physics will revolutionize ground and aerospace transportation around the world. The primary design driver for the size, weight, and cost of aerospace transportation is the need for vehicles to overcome the force of gravity to send payloads up and above the Earth's surface. The higher the payload goes, the more energy (rocket thrust) must be expended to fight against gravity and hence the larger the fuel and related propulsion infrastructure must be to carry a given payload. If an antigravity force can be harnessed, then this factor will simply go away. The antigravity transport will simply negate a portion of the Earth's gravity and propel the payload to high altitudes with minimal energy expenditure. Ground transportation could also be revolutionized by the development of antigravity vehicles that slightly negate the

Earth's surface gravity and generate above-surface hovering. The need for road improvements or new roads will disappear since one merely hovers over the surface to move.

Antigravity is strictly the opposite of gravity such that any two clumps of matter or energy will repel each other. According to Newton's theory of gravity, there is only gravitational attraction between clumps of matter and all matter attracts all other forms of matter. This is a physical law that does not depend on the electric charge or the atomic or elementary particle composition of matter, but it does depend on the distance between the two objects and the amount of mass contained within them. In the twentieth century Einstein improved upon and corrected Newton's laws of gravity and motion when he published the Theory of General Relativity. This updated theory of gravity does allow for the creation of a force of repulsion, or antigravity, between clumps of matter under special conditions.

Serious antigravity propulsion research began in the 1950s when many companies and small inventor groups were looking into anomalous couplings between electromagnetic forces and special materials or electromagnetic and gravitational forces (aka electrogravitics). Nothing came from this pursuit, so it quickly died out. R. L. Forward was the first to use Einstein's theory to describe the physics of an antigravity device that is based on the gravitational equivalent of magnetism.[7] Forward showed that to generate a 1-g antigravity field requires a device with the dimension of miles and ultradense matter flowing at ultrahigh speed within it. A resurgence of antigravity research began in the 1990s when Russian materials scientist E. Podkletnov claimed to have blocked gravity with a device based on a spinning superconducting disk.[8] The speculation is that a previously unknown complex coupling is occurring between superconductor quantum mechanics, gravity, electromagnetic, and mechanical forces to induce a gravity-shielding effect in the experiment. Since then U.S. aerospace firms, small groups, and BAe Systems, as well as Japanese, Canadian, and French firms, have worked to reproduce Podkletnov's experiment. Since 1996 the NASA Marshall Space Flight Center's advanced concepts office has spent $1.4 million to reproduce his experiment. Results have been slow in coming from any of these ventures, but they are working hard to confirm the initial results.

A potential link between superconductor quantum mechanics and gravity has been inferred from recent quantum gravity research. Another approach to modifying gravity involves the manipulation of the quantum vacuum ZPE field.[9] One proposed experiment to manipulate the ZPE involves the use of ultrahigh-intensity lasers to irradiate a magnetized vacuum.[10] If any of these are successful it will change energy issues on Earth and our relationship with the universe by allowing deep space travel.

COLD FUSION

In 1989 Stanley Pons and Martin Fleischmann at the University of Utah made a claim that shook the scientific world. Based on a claim of producing excess energy from electrolyzing palladium in $D_2O + LiOD$ electrolyte, they stated they had produced cold fusion. As their announcement did not follow accepted scientific protocols (publishing in peer-reviewed journals), the skepticism was vituperative. One of the most controversial points was the notion that fusion of any kind was being produced.

However, the announcement had such great media impact it could not be ignored. In fact, while I was at Los Alamos National Laboratory (LANL) a seminar was quickly convened, Pons and Fleischmann invited, and a discussion ensued. At LANL, as at some other labs around the country, skepticism increased when attempts to replicate the experiments failed to attain the reported results. Largely due to the uncustomary fanfare that accompanied the announcement, many physicists pronounced the whole notion of cold fusion was false and rapid decay and death followed. Or did it?

Most of those who made such pronouncements conducted quick and dirty experiments, guessing at some aspects of the protocols. Or worse, some did none at all but became convinced that since the notion of cold fusion sounded too good to be true, it probably was. However, open-minded scientists in many countries continued to be intrigued and conducted further research. Their results are largely unknown in the United States, as cold fusion experiments with positive results were virtually embargoed from standard journals.[11]

Edmund Storms, formerly at LANL and a proponent of the field, has followed the experiments quite closely. He concludes that "production of various nuclear products (radioactive and stable), radiation of various types and energy, and heat from a nonchemical source all indicate the occurrence of novel nuclear radiation. These unexpected conclusions propose a whole new field of science, a field that studies chemically assisted nuclear reactions."[12] Contrary to what most scientists think, the field of cold fusion is advancing. The scary news should be that most of the research is being done outside the United States because of the oppressive atmosphere that greets novel ideas in our scientific community.

RENEWABLE ENERGY

It would be remiss not to discuss efforts that are under way to reduce dependency on fossil fuels while reducing emissions that are causing the greenhouse effect. Many countries around the world are exploring ways to implement affordable energy while saving the planet.

It is well recognized that the technologically developed countries use energy disproportionately. Therefore, as more countries strive to improve their standards of living they will place even higher demands on the environment through increased energy production.

It is probably wrong to view renewable energy as a six sigma solution, as hard work, not breakthroughs, will be required. In reality, many significant advances have been made in the past twenty-five years and more are on the horizon. Space here will not permit exploration of the many possibilities currently being researched. The general categories of renewable sources are biomass (fuels from organic matter), geothermal, solar, wind, and hydropower. In theory, most of these resources are nearly inexhaustible as long as the Earth keeps rotating around the Sun. It has been stated that in the past 100 years people have used about one-half of the petroleum of the Earth. Whatever number you choose to describe the supply, fossil fuels have finite limits. As they are depleted, competition for them will increase.

In a single example from the National Renewable Energy Laboratory it was noted that North Dakota has sufficient wind to supply 35 percent of the electricity required in America. The issue, of course, is harnessing that power and distributing it at reasonable costs. Here, too, research has paid off. In a quarter of a century, the cost per kilowatt-hour derived from wind decreased from $0.40 to $0.05, or nearly an order of magnitude. That reduction makes wind power economically competitive with traditional power. It should be noted that wind power requires a secondary system to supplement power when the weather does not cooperate and wind does not blow continually. The necessity of redundancy is an issue to be resolved.

Hydrogen, one of the most common elements and abundantly available, offers great hope in the development of fuel cell technology. Viewed as the fuel of the future, many serious research and development efforts are in progress. The U.S. Department of Energy (DOE) has sponsored workshops focused on transitioning the country from fossil fuels to a hydrogen economy by 2030.[13] Processes are already available that combine heat and power so that production systems can double the energy efficiency.

A most important factor will be the development of efficient fuel cells that can be used to power machines, including motor vehicles. Hydrogen can be combined with other fuels to dramatically reduce unwanted emissions. An engine converted to burn pure hydrogen would produce only water as waste material. The DOE noted that "there is an emerging consensus that should global warming manifest itself, the benefits of a hydrogen economy based on clean electricity might well exceed its incremental costs."[14] Around the world more and more scientists are acknowledging that the much-debated global warming issues are of real concern. Therefore, development of emission-reduced energy is critical.

Fuel cells are under development. They work like batteries that produce electricity or heat but do not run down or require recharging as long as hydrogen is supplied. Current technologies include phosphoric acid, proton exchange membrane, solid oxide, direct-methanol, molten carbonate, alkaline, and regenerative fuel cells.[15]

Currently there are some barriers to transition to applications of hydrogen-generated energy. These include technology and engineering hurdles and the cost of the infrastructure to use hydrogen, such as containment vessels and storage media such as nanotubes or semisolid gels. Public acceptance and their concerns about safety also need to be addressed. These issues can all be overcome with sufficient resources and effort. There are no show-stoppers. Public education is required and books such as Peter Hoffmann's *Tomorrow's Energy: Hydrogen, Fuel Cells, and the Prospects for a Cleaner Planet* will help.[16]

WEATHER MODIFICATION

"It's not nice to fool Mother Nature." If you thought the experience we had with El Niño and La Niña were bad, the weather patterns over the past few years should convince anyone of the importance of weather in our lives. Extensive droughts in some areas of the world and raging floods in others demonstrate the power of the weather systems out of balance. There is a historic relationship between environmental changes and conflict, with specific examples dating to at least 1462 in the kingdom of Castile.[17]

Imagine now the ability to control weather. Once tamed, in warfare weather modification could inhibit the enemy's activities while enhancing our own. Strategically, weather modification could be used as a means to preempt armed conflict. Taken to extremes, it gives new meaning to weapons of mass destruction.

After all, the energy in a single tropical storm has been estimated as equal to 10,000 one-megaton hydrogen bombs. The superstorm Hurricane Andrew caused over $15 billion in damage to southern Dade County, Florida. The impact on people was made known to me in a most personal way. Two of my brothers lived with their families directly in the path of Andrew. As the storm veered to the south at the last minute, they had little time to pack belongings before dashing for shelter. Their homes did not blow down—they blew gone. With absolutely nothing left and nowhere to live, my brothers still annotate time as before and after Andrew.

Thoughts about use of weather have not gone unnoticed in the military. In 1996 a team under Col. Tanzy House developed a research paper that was presented to *Air Force 2025*. The title was *Weather as a Force Multiplier:* * *Owning the Weather by 2025*.[18] The paper spawned articles in many popular magazines and ultimately a formal response from Brig. Gen. Fred Lewis, Director of Weather at Headquarters, U.S. Air Force. Brigadier General Lewis noted that the Air Force had no plans to conduct weather modification.[19]

The concept paper by Colonel House, Lt. Col. James Near, Lt. Col. William Shields, Maj. David Husband, Maj. Ann Mercer, and Maj. James Pugh was quite detailed. Noting that weather was not neutral, it outlined concepts and technologies that could be used in weather modification. In degrading enemy operations they noted the following capabilities and effects:

Precipitation enhancement to flood lines of communication, reduce effectiveness of precision-guided missiles and reconnaissance, and to decrease enemy comfort and morale.
Storm enhancement to deny enemy operations.
Precipitation denial to induce drought and reduce freshwater supplies;
Space weather influence to disrupt communications and radar while disabling or destroying enemy space assets.
Fog and cloud removal to deny concealment and increase enemy vulnerability to U.S. reconnaissance and precision-guided weapons.

The study group also postulated ways in which weather modification could enhance friendly operations. These included:

---

*The term *force multiplier* is military jargon for technologies or concepts that add value to military operations. It came into vogue when the United States faced the Soviet Union and needed to *fight outnumbered and win*.

Precipitation avoidance to maintain or improve our lines of communication, maintain visibility, maintain comfort and morale.
Storm modification to help shape the battlespace environment as necessary.
Space weather influence to improve communications reliability, intercept enemy transmissions, and revitalize space assets.
Fog and cloud generation to increase friendly concealment.
Fog and cloud removal to help maintain airfield operations and enhance precision-guided weapons effectiveness.[20]

According to the report, "Current technologies that will mature over the next thirty years will offer anyone who has the necessary resources the ability to modify weather patterns and their corresponding effects, at least on a local scale."

Scientific attempts to modify local weather date back to 1946 and the discovery at General Electric Research Laboratories of a method to affect the formation of water droplets and ice crystals in clouds. Since then at least forty countries have been experimenting with cloud seeding.[21]

There are several methods for seeding clouds. They include dropping pyrotechnics on top of existing clouds, penetrating clouds with pyrotechnics and liquid generators, shooting rockets into clouds, and working from ground-based generators. Silver iodide is frequently used to cause precipitation, and effects usually are seen in about thirty minutes. Limited success has been noted in fog dispersal and improving local visibility through introduction of hygroscopic substances. This has been done over airports on occasion.[22]

More interest has been generated in this topic in recent years. In early 2002 the Division of Earth and Life Studies of the National Academy of Sciences initiated an eighteen-month study to examine technical and meteorological advances. The panel was also to identify the critical uncertainties limiting advances in weather modification. The focus is on both water management and reduction in severe weather hazards such as tornadoes and hurricanes.[23]

In 1982 I began exploring the control of weather by more novel means. Due to existing treaties the senior staff at INSCOM became very nervous. However, I was able to see firsthand the application of this system and came to believe that very substantial perturbations could be made as well as affecting local areas.

The concept was based on the research of Wilhelm Reich, a psychiatrist, once considered the most brilliant student of Freud, who developed controversial ideas about the existence of a vital energy system that he labeled Orgone. In the mid-1950s Reich ran afoul of the American medical community, and he was imprisoned until he died in 1957. While we are all aware of the stigma we attach to the book burning in Nazi Germany, most do not know that U.S. courts ordered similar measures well after World War II. Upon his conviction, the courts ordered all of Reich's papers, books, and machines destroyed. Fortunately, an underground was formed and a number of the original papers were preserved.

One of the applications of orgone energy was to modify weather. This was accomplished with a device that looked like a series of parallel tubes, grounded into a water source, which could be pointed toward the area to be affected. This gunlike device would then be used to create or dissipate clouds on command of the operator.

Trevor James Constable, a Merchant Marine communications officer who routinely sailed between Long Beach, California, and Hawaii, read about Reich's work and began to experiment with these orgone guns. Constable's duties on cargo ships permitted him a great deal of freedom in both time and action. Mounting systems on board the ships, he experimented with the process for many years. Demonstrations included filming a clear sky in all directions, then announcing that clouds would be formed off the port side of the ship. The orgone guns would be activated, and in a short time lines of clouds could be seen for miles along the port side of the ship. To starboard, the sky remained as clear as before. Constable could seemingly place the clouds in any position he desired. Importantly, he would announce the location of the clouds before activation of the system, not upon observing clouds forming.[24]

None of the aforementioned systems actually creates new weather systems. They take existing conditions and alter them slightly. Of course, that means if they create rain in one area, it does not fall somewhere else. Both intentional and unintentional consequences of weather modification should be of concern to any such operations.

The ramifications of weather modification were recognized and discussed decades ago. That resulted in the 1977 United Nations Convention on the Prohibition of Military or Any Other Use of Environmental Modification Technique (ENMOD). This prohibits the use of weather modification techniques that could result in long-lasting or severe effects.[25] As with many broadly stated treaties, the ENMOD runs counter to reality. As with previously addressed biological and chemical treaties, the research is not banned, only military applications. Similarly, it seems unlikely that potential adversaries will choose to play by the rules.

As weather is a fundamental component of the environment and could play a strategic role in future conflict, it easily falls in the six sigma category. Research in weather modification will continue for the benefit of mankind. However, it is entirely feasible that such research will be converted to military applications.

HUMAN FACTORS

Sitting attentively in the windowless room of a nondescript building at Fort Belvoir, Virginia, the agent picked up a pencil and began to write a series of brief words. That accomplished, she began to make sketches. At first they were very rough outlines. With time the drawings became more sophisticated. In the end they were nearly as good as any engineering student might produce.

The agent had been directed to mentally access a location unknown to her. Six-digit coordinates had identified the target. They were arbitrary and actually had no geographic relationship to the location. Rather, it was an assigned marker in time and space. The agent had received extensive training in an exacting process known as remote viewing. The target was an underground facility suspected of being a biological warfare facility in a hostile country. She was able to provide sufficient reliable information that the people at that facility were manufacturing biological warfare (BW) agents and preparing them for use. Based on her observations, traditional intelligence assets such as satellites, UAVs, or even human agents would begin watching for confirmation of the report. If supported, a preemptive strike would be launched and the facility destroyed before shipment of the BW agents could begin.

Already addressed was the relative importance of sensor systems. In the past decade substantial advances have been made in our technical intelligence capabilities, and they greatly contribute to our projection of military power. However, no matter how good our multispectral sensors are, they can only report what they observe at that time. They cannot predict future events except by inference and are limited to what they can directly observe.

Since the beginning of history, humans have made anecdotal references to innate abilities to foretell the future, to know what was occurring at distant locations or the status of people separated from them, and to find resources they need without any traditional means of accessing that information. In general, science has relegated these claims to unsubstantiated myths and discounted their credibility. In discrediting the field the skeptics resorted to ad hominem attacks and attempted to destroy the careers of any trained scientist who dared to explore paranormal concepts. However, a few dared to venture into the professional minefield. Their now-voluminous studies have demonstrated beyond any doubt that these nontraditional capabilities exist.

In the late 1970s Hal Puthoff and Russell Targ began a study at SRI International to explore whether these capabilities were real and could be documented. These studies, initially funded by the Central Intelligence Agency, were carried on by the U.S. military until 1995, when the program was terminated and declassified.

The name selected for the capabilities was remote viewing. This was probably a misnomer, as more senses were employed than just visual images. The official program had many names: Gondola Wish, Grill Flame, Center Lane, Sun Streak, and finally Star Gate. The research program was sufficiently successful that an operational element was formed at Fort George G. Meade, Maryland.

The first open military reference to remote viewing capabilities was written by me and published in *Military Review* in 1980.[26] In a book I coauthored a decade later I included a chapter describing how to conduct remote viewing.[27] It was heavily annotated to ensure every comment was taken from open source literature, as the program was still classified. Since the program became public knowledge a number of books have been written on the topic. In addition, a public organization has been formed to make the information widely available.[28]

The process described earlier is very close to the actual events. There were two schools of thought about the best method to attaining remote-viewing skills. One was to find people with natural ability. Certainly Joe McMoneagle was one of those. As a remote viewer with the serial number 001, Joe was known to his peers as *the best of the best*. His books recount both how he came upon the capability and how others can improve their skills.[29]

But locating people with natural ability and interest in serving the intelligence community can be difficult. Therefore, the second approach was to take individuals with military and artistic skills and train them to use their minds in unique ways. Ingo Swann made this difficult task possible through a breakthrough in understanding how information could be acquired mentally at great distances. Ingo, a talented professional artist living in the Bowery of Manhattan, had unique capabilities that came to the attention of

Puthoff and Targ at SRI. There Ingo was one of the three pioneer remote viewers responsible for providing the scientific data necessary to keep the program funded.

There were many success stories in the early research. My favorite was a session in which Ingo was sent from his controlled environment at SRI to mentally explore the planet Jupiter. Of course he did not know the assignment was extraterrestrial. Among his observations was a set of rings that were circling the planet. Since everyone knew that it was Saturn that had rings, not Jupiter, the validity of his results were questioned. Being the lovable curmudgeon and iconoclast that he is, Ingo remained adamant about his report. It was not until after March 1979 when *Voyager I* traveled close to Jupiter that it was discovered that Ingo had been right years before. Most important, that knowledge could not have been obtained by any conventional manner and ran counter to conventional scientific thought.

In addition to conducting research, Ingo was impressed to train a select cadre of the military who did not share his innate capabilities. He made a major contribution with the development of a mental model that allowed his students to acquire the desired data. Over the following years members of the detachment performed both training and operational missions employing their skills.[30]

While the program officially was declassified in 1995, as of this writing nearly 90 percent of the documents generated remain in the vaults. These include missions targeting narcoterrorists and other difficult assignments. While many of the Army's remote viewers have become known publicly, others wish to remain in the shadows. Some of them still are concerned about retribution from those they exposed.

Unfortunately, the clients from throughout the intelligence community saw remote viewing as the court of last resort. In other words, they used the techniques only when there was no other means of getting to the target. That meant that the reports could not be verified on a regular basis. The return visits by users attested to the success that the detachment enjoyed.

Joe McMoneagle demonstrated an example of the potential for strategic applications of remote viewing more than two decades ago. Early satellite coverage of a facility near the port of Severodvinsk on the White Sea close to the Arctic Circle got the attention of the intelligence community. The facility was very large, and it was obvious that some form of construction was occurring there. No conventional capability could provide answers to the question of what was happening inside the huge building. At the time, then–Lieutenant Commander Jake Stewart, assigned to the National Security Council, was a remote-viewing supporter. He decided to task the detachment at Fort Meade and see what they could come up with.

Over several days Joe McMoneagle and another remote viewer were assigned to examine the facility using only their mental skills. First Joe reported the presence of two submarines. One under development was extremely large and had features previously unknown for the Soviet Navy. He described a missile boat with twenty launching tubes and a new drive mechanism. Joe also detailed an unusual double hull and the use of special welding techniques.

None of the remote viewing material made sense, and the professional intelligence analysts scoffed at the report, noting that they would have picked up on such a dramatic change. Besides, our boatbuilders had determined that a submarine of that size would be crushed when diving. Next Joe was taken through a series of sessions in which he described how in time the Soviets would dynamite and bulldoze a path to get the sub to the sea. He even predicted when those events would occur.

On Joe's predicted schedule dynamiting and excavation began. An artificial channel was created, and the submarine was floated to the sea. When the cameras came on again, there, next to the Oscar-class attack submarine, rested a massive boat of a type that would become known as the Typhoon class. Joe had been right. The new double-hulled titanium boat beat everyone's expectations and it was not crushed as the experts had predicted.[31]

There has always been a Catch-22 associated with remote viewing. Nobody can provide an adequate scientifically acceptable theory for the means of obtaining data at a distance. There have been attempts at electromagnetic models. These suggest that there is some extremely low-frequency wave that carries the data from the target to the remote viewer. However, there are controlled experiments that demonstrate the ability to conduct accurate precognitive and retrocognitive sessions. While accepting the possibility that some wave might carry information over great distances, perturbation of time is another matter.[32]

Short a theory, remote viewing has been only marginally successful. For each success story there are failures. In addition, there are many sessions that cannot be verified. As remote viewing has now moved into the civilian sector, the best recommendation I can give to prospective students is *caveat emptor* (let the buyer beware). Remote viewing has become a cottage industry with no controls. Many of those offering training haven't a clue about what they are doing. If you hear extraordinary claims, ask to see the documentation. Contrary to what many of them say, *if they tell you, they won't have to kill you.*

There are supporters in the military who comprehend the significance of remote viewing and believe it could have strategic implications. Attending the Marine War College in 2001, Commander Rick Bremseth, a U.S. Navy SEAL with a long and distinguished career, wrote his thesis on the topic. After careful study and interviews with many of the former participants of Star Gate, he concluded that the evidence supported continued research and applications of remote viewing.[33]

Remote viewing is a six sigma solution, because it works and can radically change our means of gathering intelligence. It holds the promise of providing information about inaccessible redoubts and advances in technology. More important, once these skills are understood, those possessing them will be able to determine an adversary's intent and be predictive about events.

As with foretelling the future, mind–matter interaction has been the stuff of myths, capturing the fantasies of man by providing power over the elements. Or could there be a kernel of truth in these stories that permeate every society? In scientific terms the concept is known as psychokinesis, or PK. For many traditional scientists even the basic concept is unworthy of consideration. However, during an assignment in the Army Intelligence and Security Command (INSCOM), I had a chance to study the phenomenon fairly extensively. I and many other senior officials can attest unequivocally that the phenomenon is real. We have observed PK events firsthand and actively participated in the process.

The model for demonstrating PK metal bending is Uri Geller. While controversial, he made the phenomenon popular and attracted both acknowledgment and accusations. I have known Uri, a gregarious and affable gentleman, for several decades. Over that period I have closely observed him engaging in metal bending and have never seen him perform any sleight of hand or other magic trick. Once, in the U.S. Capitol building, Uri caused a spoon to curve upward with no force applied. In fact, the spoon continued to bend AFTER he put it down and went on with his talk.[34]

Credit for capturing the phenomenon and transitioning it into an observable form belongs to an imaginative professional aerospace engineer, Jack Houck. After more than thirty-five years working in a major aerospace company, Jack continues to provide people the opportunity to become involved in PK experimentation.

I first met Jack in 1982 when he gave a PK party at my home in Alexandria, Virginia. Among those present was my commanding general from INSCOM, Major General Bert Stubblebine. During that session we watched as Anne Gehman held up a fork by the base and had it fall over in her hand. This happened with no physical force being applied whatsoever. Totally amazed, Bert and I conferred in the following days. We determined that the implications of what we had seen were enormous and worth exploring. For the next several years I followed that trail.

The single most spectacular PK event that we observed occurred at a quarterly retreat. INSCOM had units spread all over the globe, and commanders were brought together periodically for updates and coordination. In the meantime I had learned the PK induction process from Jack Houck and was running sessions of my own. Contrary to claims of skeptics, there were notable trained observers who critically reviewed the events. Among my guests was the late Doug Henning, a magician then on a par with David Copperfield. Since Henning's manager was the first person to have bending occur that night, he knew it was not rigged. All of these people saw spontaneous metal deformation and were convinced we were on to something real.

On the night mentioned, the group of colonels and generals was invited to attend the PK event at our sequestered site. With an invitation from the commanding general, attendance was hardly optional. During the advanced phase of the PK event everyone was holding a pair of matched forks. As a slight commotion erupted at the back of the room a lieutenant colonel had one of his forks drop a full ninety degrees. Ed Speakman, a renowned science adviser, saw this happen. A colonel to the subject's right called out excitedly. Then with everyone watching, the fork straightened itself out and again bent to a right angle. Then slowly the fork continued to bend, ending up at about forty-five degrees off-center. This spontaneous bending occurred with no physical force being applied. The lieutenant colonel was a naïve subject and knew nothing of PK, let alone the sleight-of-hand tricks by which he might have fooled us. Placing the forks down, he became visibly upset. Events such as what happened to him were not in his belief system. Fortunately, we had our unit psychologist along and he was able to get the subject mentally secure enough to return home.

There has been extensive experimentation with both macro and micro PK. Jack Houck is an inveterate data collector and has records of every PK event he has ever conducted. Scientifically controlled experiments at facilities such as the Princeton Engineering Anomalies Research Laboratory conducted under the Dean Emeritus of the School of Engineering, Robert Jahn, have repeatedly demonstrated small but consistent results in mentally affecting material substances.[35]

With extremely limited funding, such research has continued. In 2001 I conducted several PK events for Dr. Gary Schwartz of the University of Arizona. Also present was Rustum Roy, founding director of the Material Science Laboratory at Penn State. Based on what he saw, Rusty stated that the reality of the phenomenon was beyond question. The next step, he suggested, was to establish a comprehensive research program in what he has termed *materials deformation.*

It is believed by many of us who have participated in PK that there is an emotional component. At a meeting on this topic at the Naval Research Laboratory in the 1980s, Bob Jahn was asked what should be of most concern involving PK. He noted that fighter pilots fly in a high-stress environment and use electronic equipment operating very close to the margin of capability and under such circumstances equipment failures are likely to be seen.

In the 1980s when I was conducting PK events for senior officials, I was frequently asked if I intended to bend tank barrels. My response was obviously no. My targets, I suggested, would be computers. As weapons systems have become more sophisticated they have come to rely heavily on information technology. Here lies the strategic vulnerability to which PK can be harnessed.

As with remote viewing, there is no scientifically acceptable theory that can account for the interaction between mind and matter. We do, however, have extensive scientific evidence of micro-PK effects. More important, the anecdotal observations of many highly qualified people offer abundant proof that these events do happen. The country or social group that first harnesses these capabilities will have a strategic advantage.

CHAPTER FIFTEEN

WINNING WORLD WAR X

In decades to come questions will arise as to when World War X (WWX) actually began. Of course, the defining moment was September 11, 2001. But as anyone in Europe or Asia can tell you, World War II commenced long before December 7, 1941. It was the sneak attack on Pearl Harbor that directly engaged the American people. Though gathering for years, the clouds of war long had been ignored. Then, so aroused by the nature of the attack, the American public precipitated a national call to arms, the battle was joined, and demarcation proclaimed.

So, too, did the collapsing World Trade Center focus the attention of all Americans on a largely unexpected threat. Enraged, they demanded a response, and the *War on Terror* was decreed. As before, this encounter with terrorism did not begin with the horrendous acts of 9/11. And as with World War II, the rest of the world had already experienced far more terrorist acts than those now perpetrated against the United States. Despite a few domestic bombings and foreign kidnappings, we had gotten off rather lightly.

Minimalist responses to acts of terrorism, drastic reduction in military strength, and emasculation of the U.S. intelligence community all contributed to the fomentation that led terrorists to believe they could successfully attack and defeat America. Potential adversaries understood the implications of Desert Storm. The effects of U.S. policy of using overwhelming force, once rhetoric, had been observed. Even a weakened American military could destroy any enemy that unwisely chose to engage in a conventional fight. Therefore, the concept of asymmetric warfare gained great favor and was studied extensively by all those who wished to oppose the West.

Concomitantly, the process of globalization, viewed by many people in developing nations as American imperialism, focused both anger and resolve. Mundane cerebral-centric thinking brought us to believe that any rational person offered increased prosperity and an easier life would naturally gravitate his lifestyle in that direction. Surprising to many Westerners, the leaders of these new revolutions were well educated and often wealthy individuals who were willing to relinquish privilege for principles. In fact, many of the suicide bombers who attacked our buildings with planes were from that class of people. Further, they had lived among us, participated in our society, and in the end rejected it in the most dramatic way imaginable.

Responding to 9/11, President Bush publicly announced a *War on Terror* and stated the resolve to see it through. He enunciated the infamous *Axis of Evil*, identifying countries that were sponsoring terrorists. Naming Iraq, Iran, and North Korea, he was attempting to define the conflict fairly narrowly and not affirm a religious component.

President Bush's efforts were met with countermeasures from Osama bin Laden, who at the time was still able to have direct access to news media. Instead of acknowledging he alone was the adversary, bin Laden called on all Muslims to participate in jihad, or holy war, against the West. In so doing he was attempting to redefine the conflict in a broader context.[1]

Whether bin Laden was successful still remains to be determined. Certainly the United States won the first round in the battle against the Taliban and al Qaeda in Afghanistan. There were a series of smaller hostilities at various flash points around the world. The full tapestry of conflict has yet to unfold. To win, the United States and its allies must defeat terrorism in detail, a feat that will be difficult to accomplish. The reality is that our first attempts have been to isolate problems and buy our way out of trouble. This will not work for the long haul.

The critical factor, yet to emerge, is whether piecemeal engagements can prevent an openly acknowledged world war. If not successfully deterred that war will devolve along religious lines. President Bush's *Axis of Evil* was inaccurately defined. There is an axis of instability that runs from Morocco to the Philippines. Already noted are the parallels: Islam versus Judaism, Islam versus Christianity, Islam versus Hinduism, and, to a lesser extent, Islam versus Buddhism. In all countries affected, Islamic fundamentalism is at the heart of the problem. Even China, with its Communist form of government, has a problem interfacing with Islamic proponents on their southwestern flank.

Despite the fundamental tenets of Islam as delivered by the Prophet Muhammad, the practice of that religion by fundamentalist Muslims has been intolerance of other belief systems.[2] While many moderate Islamic voices have been heard, they are not prevalent in most countries in which Muslims constitute the preponderance of society. Often coerced into silence, they provisionally provide support to those whom they profess to disdain. It is unlikely that the voices of reason can stem the rising tide that surges ever forward toward mass annihilation. But try we must.

The war will advance in two stages. First will come a protracted series of minor conflicts. Some of them will appear to be isolated, but they are all interrelated. During this phase the United States and its European allies repeatedly will attempt to develop coalitions. They will incorporate states that are believed to be moderate in their practices and have economic ties to the West. Some alliances will work temporarily. Others will unravel in a quagmire of self-interests. As in Afghanistan, we will learn that maintaining peace and local support is far more difficult than the military defeat of the Taliban. Eventually the tedious processes will convulse and collapse from incongruence.

In this clash of cultures, solutions that accommodate disparate values will be essential. The fundamental question that must be addressed is: Do the same values and desires motivate all humans? How you answer that question will color your perception of everything else in this chapter. There are those who answer in the affirmative; all people desire health, happiness, and an improved lot in their physical existence. This position argues that people, when presented with alternatives, will demonstrate a proclivity toward rewards most pleasing and away from those actions causing pain.

Many of those who espouse that line of reasoning also assume that democratic forms of government are inherently preferable to all others. Individuals, it is believed, should participate in governance and help determine their destiny. To that end, supporting democracies has been a stated objective of U.S. foreign policy for more than a decade and an implied concern even longer. However, in *Warrior Politics* Robert Kaplan argues persuasively that weak democracies may actually exacerbate the potential for regional instability.[3] In practice, democracy is very complex and requires experience gained through time and trial. It cannot be bestowed on emerging societies and expected to be successful. We simply must stop anticipating that other countries will act according to the American model. After more than 200 years in the democratic process, we still have a long way to go in perfecting it.

There is a substantial body of evidence that suggests that comprehensive innate values do not exist. Thomas Friedman has articulated many of these in his recent book *The Lexus and the Olive Tree*. In great detail he presents the case that many cultures are not driven by access to material things but would rather pursue an altruistic path, forgoing many physical pleasures.[4] This concept is accentuated when those deprived of adequate resources to live and blocked from any reasonable hope of advancement are offered great rewards in a transcendent life.

What was once called *the revolution of rising expectations* does not answer the circumstances with which we have been confronted. That concept was based on the notion that those deprived desire more material things. As previously noted, the terrorists on board the fateful flights of 9/11 mostly came from positions of privilege. Further, they had studied in the United States and Europe, communed freely with our society, and ultimately spurned it. Further, following those experiences in our culture and meeting our people they dedicated their lives to destroying that way of life. This should dispel the idea that "if we just get to know one another we can all get along."

The materialists' counterarguments note that young people in many areas of the world crave our music and blue jeans. They point out how quickly Western music was available on city streets in Afghanistan after the tyrannical rule of the Taliban was lifted. In many major metropolitan cities of the Middle East there are billboards hawking fast foods, Western clothing styles, and an expanding array of high-tech gadgets. Available in shopping malls, these items sell. However, for many people of the Muslim faith these are public displays of decadence. They exemplify the rationale for supporting attacks against a system so contrary to their beliefs.

Within moderate Muslim countries there is considerable potential for a dramatic shift toward fundamental interpretations. The governments of countries with existing ties to Western culture are increasingly at risk. The great disparity of wealth between the House of Saud and the general population of Saudi Arabia is but one example. A stated goal of Osama bin Laden was to bring down his own family even though they have been bribing him for years. The Saudis, with their vast petrodollars, have ensured the Palestinians stayed in refugee camps near Gaza. Through funding bombers' surviving relatives, the Saudis actively supported their suicide bombings of Israel. This overt support to the Palestinians is an attempt to appease the Saudi people. However, the now-public image of endemic corruption will probably bring them down sooner or later.

The government of Egypt is also seen by many of its citizens as a Western puppet. The people fear Mubarak is grooming his son to take power after his tenure. Historically the governments there have been very heavy-handed in suppression of dissent. The domestic economy has not thrived. A major source of income is derived from sending cheap labor abroad, not from domestic production. With only 4 percent of the land useful for agriculture and a constantly increasing population, frustration is inevitable. Critical tourism sways in the balance and ebbs and flows with each threat of conflict. Though needed to help a foundering economy, tourists offer viable targets for those wishing to further disrupt the meager financial base. The Suez stays open to Western shipping only because of the revenue it generates. Continued access is never guaranteed.

BOUNDARIES

Before undertaking the actions necessary for winning the war, it is necessary to reconsider a major imperfection in our thought processes. A fundamental flaw in American thinking is definition of boundaries. When engaged in solving any problem the first impulse is to delineate the boundaries. Most are arbitrarily assigned but quickly become treated as sacrosanct. Borders do have utility so long as they are understood as temporary demarcations, not God-ordained edicts.

A significant boundary that must be reconsidered is the basic notion of foreign and domestic. While physically boundaries are easily identifiable in some areas, they are not in others. International law is replete with such examples. Examples of the problems include definition of international boundaries at sea: Where is the shoreline? What areas have protected fishing rights? Fishing boundaries provoke fierce battles even with our close friend Canada. Arcane laws differentiate between the ownership of fish that swim freely and lobsters, which crawl along the ocean floor.

In shoreline disputes we refuse asylum to those who are turned back before reaching land but grant it to those who make it to shore. Where is the boundary in the shallow surf that recedes, leaving wet sand that will be inundated with the wave or tide? These geographic issues may keep lawyers making a profit but have little value to society.

If you doubt the impermanence of national boundaries, get out a map only a decade or two old. Locate the countries of the Balkans, members of the former Soviet Union, or those in the area then known as South Africa. Those and many other changes suggest that formal borders may fall very quickly.

Large international conglomerates span continents. Some of them have more capital than most of the countries in which they reside. Even illegal entities controlled by organized crime can destabilize geographic areas. In short, the notion of the nation-state as the fundamental building block of international relations is a useful myth. However, it is one that is waning with the realization that social organization has transitioned from a geographical base to a belief system orientation.[5] The implications of this concept exceed acknowledging nonstate actors as potential threats. It strikes at the heart of future societal relationships.

WINNING THE WAR

Winning this Byzantine conflict requires a four-pronged approach, both traditional and unconventional. However, there is also a requirement for development of more extreme contingency plans. Plan B is necessary as there is a high probability that conventional interventions will not succeed. The four prongs are:

- A strong military and intelligence community that is designed to kill terrorists and neutralize other potential threats;
- Elimination of the financial support base for terrorists;
- Energy independence; and
- Support of countries and non-governmental organizations that weaken the disparate fundamental socioeconomic issues that foster anti-American activities.

STRENGTHEN THE MILITARY

While Secretary of Defense Rumsfeld has begun a process of military transformation, it falls short of what will be necessary to ensure our security. What is needed is the creation of a Department of National Security (DNS). This is not a renamed Department of Defense. Just as the Department of Defense evolved from the War Department and became very different, so, too, would a Department of National Security be substantially different from its predecessor. It should be responsible for all security issues without regard for origin of the threat. It must have the ability to function across agency lines in ways that coordination will never accomplish. The basic premise is addressed in the oath of the military that states: "I will protect and defend the Constitution of the United States against all enemies, foreign and domestic."

On the one hand, too many senior bureaucrats suffer from the inability to comprehend the obvious. Politicians, on the other hand, refuse to act unless their reelection self-interests are met. Between these two groups the public suffers inane responses to easily solvable problems. Worse is when bureaucrats and/or politicians take measures to prevent competent people from addressing the issues. As an example, in Yemen Ambassador Barbara Bodine blocked the reentry of John O'Neill, the FBI's top counterterrorism expert, who was investigating the bombing of the USS *Cole*. Despite the massive blast and a viper's nest environment, she questioned why the agents were armed.[6] Of more concern should have been that then FBI Director Louis Freeh, a consummate, albeit unimaginative, bureaucrat, supported her actions.

For years the palpable warning signs of the rising epidemic of terrorism were largely ignored at the top. A series of events, culminating with 9/11, finally made crystal clear that the American mainland was at risk. After months of bickering, on November 20, 2002, Congress finally passed a bill authorizing the biggest reorganization of the Executive Branch since 1947. With action sorely needed, the bill became bloated with pork and enmeshed in backroom deals as politics as usual again took precedence over our security.

Creation of the Homeland Security Department was a politically acceptable attempt at solving the serious problems associated with coordinating activities of disparate agencies. Unfortunately, it does not get at the heart of the issue—the national security. The organizational response leaves the Department of Defense intact while congealing other agencies that have responsibilities pertaining to domestic tranquillity. The enemy is neither split nor mindful of our self-imposed boundaries. Therefore, we should develop a unified organization that meets today's requirements. The old system is anachronistic.

This proposal was first made in a presentation to the Defense Science Board in 1994.[7] Examination of existing threats suggested that our asymmetric adversaries have studied our institutional construction and looked for weaknesses. In football parlance, this presentation predicted that the enemy would run the seams. Locating the boundaries of organizational responsibilities, enemies would strike at areas of overlap or underlap. The near ubiquitous interagency rivalry and bickering is well known as, under the *Golden Rule*, they compete for their budgets. Recent reviews of intelligence failures prior to 9/11 strongly reflect these predictable problems. While coordination has improved substantially, when faced by complex threats, a single leader is an absolute necessity.

Missing within the existing institutional framework is a comprehensive strategy for both national and international security. This issue provides banner headlines across the globe. My private conversations reflect the same thing. Well-educated people from various countries indicate that they simply cannot figure out what U.S. policy is. The inability of anybody to articulate our goals fosters an environment in which rumors run rampant. Allies and adversaries alike are befuddled, but astute enemies can quickly seize the perception initiative and force us into a reactive mode. Since the end of the Cold War we have been in a reactive mode.

Strategy implies long-term commitments to goals. History has proven that the United States is rarely a reliable ally. We have stayed with NATO and Korea. But there is also Vietnam, Laos, Lebanon, Somalia, and even Iraq. Remember when we asked the Kurds to rise up? They did and we didn't. Then, when key people were brought to the United States for safe haven, immigration authorities put them in jail for years. This is not the first time people have responded to our suggestions and been left holding the bag. Older readers may remember the Hungarian revolution that began in October 1956 when they attempted to oust the Soviets. They were responding in part to the broadcasts of Radio Free Europe that strongly implied help would be provided. Their pleas went unheeded and the uprising was brutally crushed.

Afghanistan serves as another example. When the *mujahidin* were fighting the Soviets, we befriended them. As soon as the Soviet threat dissipated, we pulled up stakes and left. Pakistan hasn't forgotten how we used their territory for that operation. They, too, were forgotten. Enduring Freedom was a mess of our own creation. Few people are aware of the secret war we supported in Tibet. Then there was the Ethiopia/Somalia fiasco in which we switched sides based on what the Soviets were doing in the area. If you are in a Third World country and you hear someone say, "Hi, we're from America and we're here to help you," be careful.

Developing a strategy is essential in tailoring the future military. In a direct confrontation we can defeat any traditional adversary. But what the military calls op tempo, for operational tempo, or the level of activity, is the critical factor. Fewer troops are being asked to do more—continuously. There simply are not enough troops to disperse them around the world on long-term commitments.

Defeating the Taliban in Afghanistan took a few months. Holding the area will take much longer, possibly decades. For years many troops have been stationed in the Balkan countries, and there is no end in sight. Annually Special Operations forces are deployed in more than 150 countries. Continuing at that pace places monumental stress on those with families. As a consequence, divorce rates skyrocket. The option is for experienced people to leave, and they are very hard to replace.

To meet operational requirements reserve forces are being activated on a routine basis. They have day jobs. When called to active duty they stop supporting their local economies. The day is gone when they could be called *Weekend Warriors*. Additionally we have moved an increased number of combat arms forces (the people who fight) into the reserve components. Modern combat is too complex to be done on a part-time basis. Constant training is necessary or the price will be in lives. At present, our ground forces are just too small to accomplish the tasks assigned them. The answer is to rebuild the light infantry brigades and divisions, as they will be needed in the near future. In fact, if the current op tempo is to continue, we will need to increase forces for all services. And there is every indication that commitments are going to increase.

We must stop viewing terrorism as if it were a street crime. Terrorism is inherently different. While the perpetrators may be thugs, they are bent on our destruction. Lawyers are already too involved in the battlefield. If they want to participate they should enter the crucible, literally not figuratively. Once a political decision is made that we are going to war, it must be understood that collateral casualties may occur. The ruling that positive identification of specific individuals must be made before an attack can be initiated is ludicrous.

Yet lawyers did require that before missiles could be launched from Predator; the optical targeting had to be equal to that of a soldier at 100 meters with binoculars. In October 2001 Navy Captain Shelly Young, a U.S. Central Command lawyer, advised against launching a missile against a vehicle because women and children might be in the convoy. That recommendation probably let Mullah Omar escape. If we are going to fight terrorists, we must outline the rules of engagement and get the lawyers out of the command center. If you don't want to risk killing people, then don't go to war.

The technology base has been used as a bill payer for too long. The public doesn't comprehend the complexities of defense budgeting. However, they should know that there is no line item called *go to war*. That means that for each of the myriad operations in which they engage, something has to pay for it. Training and research have been shortchanged for too long. It is time to stop eating the seed corn and invest heavily in basic research.

Similarly, the intelligence community has been nearly decimated. Disgusted with lack of support from the White House and emphasis on politically pleasing intelligence reports during the Clinton Administration, many experienced professionals left. The intelligence community has never fully recovered from the debacle of Admiral Stansfield Turner. Under President

Jimmy Carter, Admiral Turner destroyed our extensive human intelligence capability in favor of spy satellites. Time and again we have been caught short because we lacked the ability to have human agents inside the enemy's camp. Sometimes the people who spy for you are not nice people. But after all, we do ask them to betray their allegiances.

In the complex world of terrorism, spies will be even more important. Developing such nets takes a long time, and trust is essential. The revelations of moles in the CIA, the FBI, and other agencies will make potential agents very leery of supporting us. After all, it is their lives that are on the line. This does not diminish the role of the National Reconnaissance Organization or any of the other technical collection means.

An inherent problem that emerges in the use of human intelligence in counterterrorism involves the previously mentioned boundary between foreign and domestic. In the coming war, part of the enemy will be us (American citizens). Splitting domestic and foreign intelligence is fraught with danger. That can no longer be doubted. However, the individual rights versus need for security will constitute a Gordian knot. There are legitimate issues that support protecting rights. There are also strong arguments favoring increased security. These battles will go on in court. Unfortunately, it is at this seam that our adversaries will attempt to stop observation of their activities. Extensive court battles must be anticipated. The amount of pain inflicted on the United States will impact the relative degree to which the courts support one side or another. However, the courts must be perceived as effective or citizens will resort to direct confrontation.

For years discussion about new nuclear weapons has been politically incorrect. In fact, even discussing testing programs for stewardship was verboten. The *nuclear threshold* is another arbitrary boundary that was established decades ago. Since then conventional weapons have been developed that have enhanced energetic capability. Also, nuclear systems have been miniaturized and long-range precision dramatically improved. There exist a class of targets for which small nuclear weapons are the best solution. These include deeply buried command bunkers and other underground critical facilities. We should maintain an arsenal of earth-penetrating nuclear weapons that can hold such targets at risk. It should also be made known that we will employ these weapons to take out those facilities if they threaten the United States.

No matter what happens, Special Operations forces (SOF) will play a major role in future operations. Although these units are expensive to train and maintain, extra resources should be provided to them. This will be necessary to keep them deployed around the world. These soldier diplomats are crucial in establishing personal relationships and having our eyes and ears in places satellites cannot reach.

There is a need to change the personnel system so that SOF members can be used at any stage of their life. With access being a high priority, they should take advantage of people who have worked in foreign areas or travel routinely. Development of a discontinuous pattern of service would help. This is another boundary falling, as God does not require that service members serve all, or even most, of their term consecutively. There should be options to easily obtain the services of those with special expertise, knowledge, or personal relationships, even well into retirement.

With rising crime and violence levels in many areas of the world, private security companies are rapidly increasing. Some are very professionally run and routinely hire former soldiers with special operations backgrounds. Just as the government supports the CRAF program for airplanes, it could establish relations with accredited security companies. The companies would be provided money and equipment. In return, they could provide personnel with requisite skills when they are needed for specific missions. The people would be sharp, as they would use their skills in their daily lives and yet keep up to speed with military requirements and clearances as necessary.

This is not the same as outsourcing wars. There would be quality control and a much-needed capability to rapidly increase SOF operations on demand. Unlike reservists, the personnel involved would exercise many of their skills on a daily basis. They would return with language proficiency and area orientation in place. Very little training would be necessary prior to deployment. An innovative move would be to afford the SOF augmentation personnel periodic vacations in their areas of interest so they can renew friendships and keep current on changes in the area.

Information dominance is an absolute necessity. Units have been formed to conduct both defensive and offensive information warfare operations. The most significant problem will be securing the critical infrastructure, including our telecommunications and financial system. Despite increased emphasis in this area, the current organization is probably insufficient to insure our security. This is an example of where the Department of National Security would provide the overarching structure necessary.

Non-lethal weapons will become more important in future conflicts. At least in the initial phases we will face many situations in which terrorists are in close proximity to noncombatants. There is a need for area non-lethal weapons. High on the priority list should be acceleration of the development of antipersonnel incapacitating or calmative agents. We need the ability to separate terrorists from hostages and other non-combatants. The research was stopped decades ago for the wrong reasons. These weapons will be critical when battles move to London, Paris, Rome, Washington, New York, and Los Angeles. Failure to develop these incapacitating systems will ensure that lethal force is used when non-lethal alternatives could have been available.

The Russian experience in freeing hostages at a Moscow theater in October 2002 proves the technology can be made available. From prior attacks, the willingness of the Chechen terrorists to kill their captives was well established. The vast majority of the accidental deaths were derived from tactical and judgmental errors. Had doctors been alerted to expect victims to be treated as if they had heroin overdoses, the doctors would have saved many. In addition, adequate personnel to keep each person breathing would have ensured that even more survived. However, before becoming too critical of the Russian response, remember, it was the FBI that kept fire engines away from the burning buildings at the Branch Davidian compound in Waco.

If it is believed that these non-lethal weapons might violate treaties, then alter the legal parameters. The old treaties were drawn up for other purposes. Those who are persuaded that such treaties *are the bulwark of our chemical and biological defense* should listen carefully to the words of Ken Alibek. After rising to second in command, he defected from Russia and advised the West on the most comprehensive biological weapons program the world has ever known. It was initiated the same year that the United States and Soviet Union signed the chemical and biological warfare treaty. Those who argue that we should rely on these treaties are placing our security on a foundation of quicksand by delivering false hope. My admonition is simple: We have the wrong laws; we must change them.

ELIMINATE FINANCIAL SUPPORT FOR TERRORISTS

The primary goal in combating terrorism should be to eliminate their financial base. The process to do that was articulated in the chapter called "The Root of All Evil." Each proposal listed there could be implemented. For reasons misguided, most of them will languish. The result will be cyclical terrorism and eventually war.

ENCOURAGE ENERGY INDEPENDENCE

The asymmetric response to major conflict is energy independence. "Six sigma solutions" suggested a number of technologies that could meet our requirements. There is a direct linkage between energy independence and national security. A hydrogen economy is well within our grasp. No major breakthroughs are needed save convincing the American people of the strategic necessity for the change. The rest is hard work. In addition, other renewable energy sources would make substantial contributions. They include wind, solar, and hydropower for electrical generation. Geothermal and biomass technologies are not far behind.

There should be aggressive funding of high-risk, high-payoff energy systems. We can no longer afford the *not invented here* approach of bureaucrats and other turf-protecting scientists who spontaneously reject any new technology they don't understand. There is a big difference between technologies that have a theoretical basis and naïveté. This is an area in which we dare not let some other nation or group make a breakthrough before us. Examples of technologies include fusion, which has a high probability of success, zero point energy, antigravity, and cold fusion, each of which does have the risk of successful application associated with it. The stakes are simply too high to ignore.

A HELPING HAND

In an attempt to forestall large-scale conflict we must attempt to provide support to countries and non-governmental organizations that weaken the disparate fundamental socioeconomic issues that foster anti-American activities. While most Americans believe the government spends a large amount in foreign aid, it is really minuscule when compared to the size of the budget. Though we spend about $11.4 billion, this represents only 0.01 percent of our GNP. Comparatively, it is the smallest amount any of the technologically developed countries spend. Even tiny Netherlands with 5.3 million people contributed $3.2 billion in 2001.[8] To equal that comparative amount the United States would have to increase foreign aid about fifteen times the current rates.

There is an obvious need to increase foreign aid to countries that are likely to produce the adversaries of the future. As uncontrolled population increase is fueling much of the economic disparity and producing the future generations of terrorists, birth control should be at the top of the priority list. It is absolutely essential that we support all methods of birth control in developing nations. The *holier-than-thou* politically and religiously motivated anti–birth control restraints on foreign aid must stop. This position literally will kill us in the long run. The paradox is that those who refuse to support birth control are perfectly willing to kill those children when they become adults and threaten us. It is strategically important to the United States that poor women throughout the world have access to cheap and effective birth control measures.

While education is not a panacea for countering terrorism it will help ease problems in many areas. Few Americans can begin to comprehend the true status of education in rural areas of the world. Traveling in Nepal, which has only a 50 percent literacy rate, provided me a clear picture. While visiting schools, we found crowded facilities with almost no supplies. Paper and pencils were gratefully received. Students were chanting by rote and calling it reading. Few would stay past third grade, as they were needed at home for subsistence farming. Therefore, it came as no surprise when I found indications of Maoist guerrilla activities close at hand in all of these areas.

Unfortunately, this Nepalese experience is typical of underdeveloped countries. There is no comparison between what Americans learn in school at a given level and what these deprived people have for education. For limited cost we could provide both teachers and supplies that would dramatically improve these educational systems. Expansions of these programs should be a national priority.

Whenever money is dispersed, the potential for corruption is present. One of the inhibiting factors for support for foreign aid is the horror stories concerning the large amounts of funds that are diverted into private coffers. Along with increased funding must be clearly delineated controls with severe penalties for misappropriation.

Perception management is considered a dirty word by many politically correct bureaucrats. They are wrong. When the Department of Defense opened such an office under Brigadier General Pete Worden there was a spontaneous hue and cry from defenders of civil liberties. The office was publicly executed in short order. That was a mistake.

As part of our efforts to counter emerging adversaries we should engage in an unapologetic effort in global perception management. How others see us will be the single most important factor in future conflicts. It determines who is with us and who is against us. Negative opinions provide fertile grounds for recruitment of enemy forces. Our image is too important to be left to others to portray. Competing opinions will also be provided. That is

fine in a democracy. However, the national-level program to influence opinions can and should be developed and implemented.

PLAN B (WHEN THE PRECEDING STEPS FAIL)

There is no guarantee that we will be successful in preventing escalation of
the *War on Terrorism* into a much broader war. While we would choose
otherwise, it may well be thrust upon us. For the reasons previously stated,
there is a strong probability that the United States may have to go it alone.
While we would prefer to form alliances, we cannot depend upon them in
all circumstances.

Therefore, we must do the following. First, we must have the military
capability to defeat any adversary. Most threats can be defeated with conventional forces supplied with advanced weapons systems. However, given
our relative size and strength, that strongly suggests we resist unacceptable
or unilateral cuts in our nuclear arsenal. Safeguards should be initiated to
ensure that the weapons will work as advertised. That may mean periodic
underground testing.

A major portion of this proposal appeared in the chapter called "Rethinking Space Missions." It runs counter to current space treaties. But at this
moment in time we have the capability to implement extensive programs
that do include weaponization of space. This does not mean exclusive use
of space but does provide the U.S. control of that domain. Undoubtedly
we will welcome any peaceful enterprise that wishes to operate there. We
will just control the defensive aspects. The ability to develop this option is
time-sensitive. Our great reduction in space research and technology budgets
has placed our lead in uncertainty. Other countries, including China, are
rapidly expanding their capabilities for space exploration.

We know how to get to the Moon and beyond. We have not done so
recently as a matter of political will, not technical expertise. Placing a permanent base on the backside of the Moon makes sense for military and
planetary defense reasons. It would serve as a joint command and control
center and sensing outpost with multiple applications.

There are those who argue that there is no need for manned space and
that dominance can be achieved from terrestrial bases. That issue remains
to be resolved. It would run counter to man's traditional drive to explore
the unknown.

Pax Americana? A decade ago I would have considered that unthinkable.

Ralph Peters and others have noted that America needs to understand and embrace its imperial role. This is not a mantle easily accepted by our philosophy. Yet it is being thrust forward. In my view we are not yet ready to take the measures suggested in Plan B. It is, however, time for someone to begin the contingency planning. Imperialism, with whatever dangers lie therein, may be required.

However long we delay in taking the steps necessary to defeat terrorism will directly influence how drastic our later actions must be. How harshly we respond will be related to the amount of pain suffered by the American people. We have been told that 9/11 was to be the first of three attacks, each escalating in ferocity. Any further attack now would likely be met with a response so unmitigated that no adversary would dare launch another.

Allowed to gestate, these problems will mushroom and we will be faced with a complex world war based on belief systems, not geography. The actions necessary to win that conflict are nearly unthinkable today. While it is not the future we want, it is one we had better be prepared to win. Survival of our culture is at stake. Paradoxically, our culture will have to change in order to predominate. However, as we have repeatedly demonstrated, when challenged, the American people can and will rise to the occasion.

My message to those who mean us ill will is this: *Tread lightly, for we are indeed a sleeping giant.* Our slumber has been interrupted and tempers are short. Should great harm befall us, you cannot imagine the conflagration that will follow. Asymmetric or not, in out-and-out war with the United States there is but one rule: WE WIN!

May you live in interesting times.
—AN ANCIENT CHINESE CURSE

ANY BOOK DESCRIBING FUTURE events is fraught with risks associated with unforeseen but important shifts between the time the manuscript is written and the publication date. Thus it was with *Winning the War*. We live in a turbulent world replete with the mutating social structures described in Appendix A of *Future War*. Often unpredictable in nature, they cloud our ability to forecast events. The significance of the nation-state continues to devolve, though it is a concept tenaciously embraced by those appointed to existing positions of power. In addition to the complexity of international relations and decisions concerning use of force, sometimes accidents occur that play significant roles in our future.

MANNED SPACE

On February 1, 2003, the telltale multiple white streaks across the azure Texas sky left an indelible mark on America's manned space program. The scheduled reentry of the space shuttle *Columbia* went terribly wrong. Catastrophic failure of the heat shield brought about rapid disintegration of the craft, scattering debris over several states. With that accident came yet another pause in American exploration of space as experts diligently examined sensor recordings and tiny scraps of physical evidence in an effort to determine the root cause of the accident.

The destruction during the January 28, 1986, launch of *Challenger* stalled NASA's shuttle program for more than two years. This time, however, people are stationed aboard the International Space Station, and it is up to the United States or Russia to get them back. There are plenty of supplies and a rescue craft that can be used in a last ditch attempt to return the three-person crew back to Earth.

Columbia's fate brought forth renewed questions concerning the efficacy of manned space missions. Exceedingly expensive and inherently dangerous, many experts have argued that space exploration should be relegated to remote sensors and robots. However, in chapter 13, "Rethinking Space Missions," I have opined that commercialization of space would dramatically increase manned presence in that harsh environment. I still believe that to be the case. The *Columbia* investigation will provide a short, albeit necessary, respite in our manned efforts. The age of the fleet and prior budget constraints will be questioned. Then, within a short period of time, we will continue our manned exploration, for our destiny is in the stars.

Economic interests will drive the coming space programs. No mode of transportation is without risk, and industrial mishaps occur in all fields of endeavor. The unforgiving domain of space will see many more casualties. It is highly likely that the next accident in space will be in a commercial craft, not one of government design. Despite the dangers, humans will venture forth. Unsung heroes, like the nearly seven hundred brave souls who previously have slipped the bonds of Earth, will continue to lead the way. There will be more eulogies that follow the somber condolences and prayers for the crew of *Columbia*. But, the quest will continue.

SHIFTING SANDS

The end game of space is likely to be determined by terrestrial events. Preparations for a war with Iraq brought into doubt the viability of long established alliances. Relations with the organizations that formed the very foundation of international security and cooperation were severely strained. On February 5, 2003, Secretary of State Colin Powell presented to the United Nations Assembly irrefutable evidence that Saddam Hussein had not complied with UN Resolution 1441. For more than a decade Saddam had dodged and weaved his way past weapons inspections and failed to prove that he had destroyed his vast quantities of chemical and biological agents as required under the 1991 UN Resolution 687; yet, many countries still refused to authorize further action and called for increased inspections. Several of our traditional NATO allies dug in their heels in order to protect their shadowy business interests in the Middle East.

France, suffering from withdrawal symptoms caused by dramatically diminished power that brought the country to near irrelevance, behaved obstreperously. However, their permanent position on the UN Security Council provided a forum for pompous posturing. Of course, they were not alone in opposing a war with Iraq. Russia, walking a fine line between old and new partners, also resisted military action against their former client state. China, another permanent member, chose not to support action.

While leaders of many other countries came forward to support the move to disarm Saddam, they faced tremendous opposition from their constituents. The largest peace demonstrations in the history of Europe took place on February 15, 2003. More than a million marchers were reported in Rome, while other key major metropolitan areas had huge crowds. Reminiscent of the Vietnam era, across the United States throngs took to the streets to indicate they were against the impending action. So great was the global outcry that President Bush slightly moderated his call for action.

Instability on the Korean peninsula precipitated dramatically when North Korea announced it was reopening its nuclear plants. They ordered UN inspectors out of the country and made bellicose statements threatening the U.S. mainland. Kim Jung II stated he could place any part of the United States at risk, as they had nuclear weapons and long-range missiles to deliver them. Upping the rhetoric, using the state-run paper *Rodung Simmun*, II, on his sixty-first birthday, beseeched all Koreans to, "burn with hatred for the American imperialists."

Despite stated urgency in disarming Iraq of weapons of mass destruction, threats from North Korea were greeted with moderate statements indicating the situation should be handled diplomatically. This mild mannered approach seemed incongruous to many observers, especially when China had little success in toning down the rhetoric from Pyongyang.

Concurrently, a major rift appeared between South Korea and America that certainly threatens the future of the area. Roh Moo-hyun, the new president of South Korea, was elected on an anti-American platform. There have been many demonstrations and acts of aggression against U.S. troops stationed in that country. LTG Charles Campbell, the U.S. Eighth Army Commanding General stationed in Korea, was literally brought to tears on *60 Minutes* as he described his feelings about seeing the American flag burned by people whom he was there to defend. In the same program, Korean civilians in Seoul stated they feared President Bush more than Kim Jung II. That was amazing, as the latter was poised with a million-man army within artillery distance from them. Recently, II has expanded his special operations units to 100,000 men designated to disrupt their country. Despite those facts and that we have spent tens of billions of dollars to insure the sovereignty of South Korea, the younger generation has forgotten the past. They have no concept of the clear and present danger facing them. In their zeal for reunification, they seem to have developed a misguided notion that the Dear Leader is a reasonable man who will accommodate democracy.

TO WAR OR NOT TO WAR

Ambiguous threats can be just as deadly as overt ones. The convenient method of determining when a country would go to war based on simple rules of good and bad is long extinct. For centuries, territorial invasions have met with armed resistance. While conflicts were often driven by ulterior motives, there was normally a critical event that precipitated action. As has been demonstrated in several recent applications of force by the United States, the old system has changed. Now there is a *calculus of conflict decision-making*. This calculus integrates many complex functions and rarely rests on a single event. Often, a series of small incidents builds to constitute an intolerable situation. Thus was the case for the actions in Grenada, Panama, Haiti, and Somalia. On occasion, it takes several major events before we act. Even genocide, such as in Rwanda or the ethnic cleansing of Bosnia, did not acquire prompt attention. In all of these situations there was a previously undetermined point at which the decision was made to use force.

The continued proliferation of weapons of mass destruction further complicates the calculus of conflict decision-making. Even their potential availability or use may dictate the necessity for preemptive strikes. The rationale for timing and method of reducing the threat will be cloudy at best. In fact, a degree of ambiguity concerning our decision-making strategy may act in our favor, provided the adversary is convinced that we have the *capability, intent*, and *will* to destroy him. However, these same factors do not play well with the media, or often our allies.

The lesson is that Americans, and others, must come to understand the difficulty of making these decisions. They are not sound-bite friendly. It will take a well educated populace to understand the full parameters of the issues. Certainly there will be room for discussion and dissent, and rarely will we experience unanimity. That's the American way. However, as a country we should determine our tolerance thresholds concerning use of force in defense of our national interests. Once those established lines have been crossed, the decision to employ force must be unilateral. A sure road to defeat is combat by consensus. Concomitantly, the proliferation of weapons of mass destruction demands preemption, not retaliation for attacks.

In many countries there are cavernous differences of opinion between the general populous and their leaders concerning use of force. This is indicative of serious problems in the near term. The points of contention are brought about by a fundamental lack of adequate education and understanding of the complex problems that we will face in the future. Classically, we have been taught to believe that conflicts are clearly defined between *good* and *evil*. In fact, our leaders have accentuated that notion. President Reagan called the Soviet Union an *evil empire*, while President Bush the younger coined the phrase *axis of evil*. Such elementary notions may be useful for rallying public sentiment but fail to portray the intricacies of the real world. The flaw in such thinking, and one that is potentially fatal, is that they offer single point failure should any aspect of the concept be exposed as untrue. Still, it well fits the mold of sound bite philosophy and our television script mentality that reduces complexity to absurd simplicity and any problem solvable in an hour.

This is an anachronistic viewpoint about threats to our security. Though it is held by most Americans, it needs to be revised. The values conjured up in images of the Old West no longer apply in modern war. The sheriff meeting an outlaw in the middle of the street at high noon and waiting for him to draw first just won't work. Given the change in the nature of warfare, tit for tat responses predicated on the adversary attacking first are likely to be suicidal. Technologies for weapons of mass destruction have moved too far to risk absorbing the first blow and then responding with measured force. In recent months, President Bush has made it clear that our policy of no first strike is obsolete. This has correctly put potential adversaries on notice that evidence of a serious threat may precipitate a preemptive strike.

In wars past, we have often played catch-up. During 1914, when World War I broke out in Europe, the United States had less than 100,000 men under arms. In May 1915, our merchant ship *Gulflight* was torpedoed. A week later the British liner *Lusitania* was sunk off the coast of Ireland, resulting in 1,198 lives lost. Of those, 128 were American citizens, resulting in a tremendous public outcry and indignation. However, it was not until 1917, after many ships had been sent to the ocean floor, that the United States finally raised an Army and formally entered the war.

In 1941, again years after the onset of conflict in other areas of the world, the Japanese attacked Pearl Harbor. That strike nearly crippled our Pacific fleet, and only the forward thinking of Admiral William Halsey, who already had his ships at sea, prevented all of our aircraft carriers from certain destruction. Even then, it took two years to fully regroup our forces and begin to fight our way back. In reality, the United States had consciously stayed out of the fray and had taken the intervening years to retool its defense industry.

The beginning of the Korean War in June 1950 saw the Republic of Korea Army, supported by inadequate U.S. forces, retreating to the Pusan perimeter. There, they barely held on until the U.S. invasion at Inchon broke the back of the North Korean Army months later. Vietnam proved the need to reconfigure forces designed to wage an armored war in Central Europe into jungle fighters. Desert Shield began after Iraqi troops had completely overrun Kuwait. Before Desert Storm could be launched, it took six months of concerted effort to place both troops and logistics in theater. Then, active duty troop strength was high enough to execute the mission with limited call-up of reserve forces. The drastic military cuts taken during the Clinton administration have meant that nearly every application of force requires extensive activation of reserve units. Activation of those troops takes time and detracts from the civilian economy. Soldiers and employers will tolerate and support our troops when our national security is overtly threatened, but repeated call-ups for unclear causes severely stresses everyone involved. As a senior reservist told me, "Our people vote with their feet."

The bottom line will be whether the United States chooses to maintain the flexibility to operate unilaterally and decisively or agrees to enter into unending arcane debates that will be constantly held hostage to the political whims of third-rate powers. It will be better to have coalition support before engaging in conflict. Consensus for multilateral support, however, will be attained more easily provided it is clearly understood that we have the necessary force to accomplish the mission without it. That power will limit the political gamesmanship and focus debate on issues of substance. Therefore, it is likely that the United States will opt to dominate space while we still have the technical advantages necessary.

Great power brings great responsibility. Many countries, including some former allies, will resist a U.S. move toward domination of space. We will be accused of being a global bully. Therefore, we must avoid indiscretion in use of force. Should we fail to take bold steps, survival of our nation as we know it may be at stake. We can, and must, ensure that friendly nations are adequately provided for in their access to space for peaceful purposes. However, control of the *High Frontier* is the most reliable method to ensure rapid and decisive response to threats emerging anywhere in the world.

ADAPTATION TO THE *WAR ON TERROR*

In the few intervening months since this book was drafted, realities of the *War on Terror* have brought about substantial changes in the United States Government's response. While these actions are well short of what is needed, they probably press the limits of political viability at this time. We simply have not been hurt badly enough for the American people to get serious about the new realities of life.

HOMELAND SECURITY

Possibly the most visible response to the *War on Terror* was creation of the Department of Homeland Security and appointment of the first secretary of that office, Tom Ridge. Combining twenty-two organizations under a single new department, Homeland Security set about a near hopeless task. Impossible expectations were set by Congress, the public, and the media while the most fundamental issue was largely ignored—the Golden Rule. Congress simply does not appropriate funds in a manner consistent with rapid reorganization. Many companies offered goods and services to the new agency. Despite the ubiquitous promise of large pots of money, almost none began flowing as bureaucratic entanglements, lack of personnel, and pure politics bogged the system down.

One of the first publicized actions was to announce the Homeland Security Advisory System. When presented to the public, the overly simplistic five-color threat level chart was greeted as the butt of many jokes. Basically, people questioned whether a system that shouldn't have taken a manager over an hour to construct was the best that could be expected from this highly acclaimed office. Before long, however, a change in color could affect the stock market, even if the average person just didn't get it.

The other shortfall was staying out of the rice bowls of the 800-pound gorillas. The Department of Defense formed the U.S. Northern Command, while the Intelligence Community remained pretty much intact. Their relationship with Homeland Security will be one of coordination. While this may be politically correct at the moment, it must be revisited at a later date.

LEGAL ADJUSTMENTS

The Patriot Act, HR 3162, signed into law by President Bush on October 26, 2001, laid the framework for the war on terror. That sweeping act provided law enforcement and intelligence officials a new set of tools. It provided the ability to place surveillance on terrorists, unprecedented sharing of intelligence, and to have special judges handle sensitive matters. Importantly, it beefed up response to the money laundering procedures previously mentioned. Those concerned about security hailed HR 3162 as a godsend for their protection. Most civil libertarians decried the act as an attack on the foundations of our democracy.

In the following period, surveillance aspects of the act were substantiated when tested in court. Terrorists were detained and their rights abridged. President Bush pressured normally feuding intelligence organizations to cooperate in an unprecedented manner. But more may be needed. Early in 2003, the Justice Department draft of a sequel to the Patriot Act surfaced. Various names were attached to the draft, including *Patriot Act II* and *Domestic Security Enhancement Act of 2003*.

Provisions of the new act might include greater restriction on information available through the Freedom of Information Act (FOIA). Some of the areas further protected from FOIA include sensitive business information and suspected terrorists in custody. They recommend establishing a DNA database for terrorists so that even small pieces of bodies can be positively identified. That would be helpful in cases of suicide bombings in which most identifying features are destroyed. Other controversial points would include termination of all state law enforcement consent decrees issued before 9/11, establishing presumption for pretrial confinement for terrorists, and expatriation procedures for Americans who engage in terrorist activities. It remains to be seen whether this act will be introduced and passed. As incidents of terror increase, this, or a similar bill, will become law.

A significant problem in addressing terrorism and other use-of-force issues is that most Congressional leaders are inexperienced in those areas. The time has long passed when the majority of Congressmen had served in the military at some point in their lives. Now very few have, and they lack the firm understanding of what it means to serve in the military. No number of briefings or tours of operational areas or bases will provide the depth of knowledge that comes from serving. Worse, just like many of their constituents, they want simple solutions to complex problems.

This was graphically demonstrated when some Congressional leaders questioned whether we were giving up the search for Osama bin Laden to attack Saddam Hussein. I assert that the War on Terror is but a precursor to World War X and they are not separate and distinct. In reality, issues of national security are integrally linked. When our Congressional leaders can't understand these complex issues, how can we expect the citizens to comprehend them?

ASSASSINATION

One example of a more aggressive policy is that assassination has been publicly announced for use on a slightly expanded scale. An Executive Order was signed following 9/11 that targeted approximately two dozen terrorists whom officials called "the worst of the worst." The most high profile of these attempts was the use of an armed Predator in Yemen. In November 2002, an antitank missile killed Qaed Salim Sina al-Harethi. Five other terrorists died while riding in the car at the time of the attack. One of those was an American.

The current process for authorizing assassination is, and should remain, highly formalized. However, guidelines need to be preestablished informing operatives under which conditions a strike can be made. Included in the guidelines should be the level of collateral casualties that are acceptable before a weapon, such as a missile, can be employed. The rules should be simple. Lawyers may be involved in drafting and reviewing the regulations but should not be placed in operational decision-making positions. Agents cannot be second-guessing as to whether their authorized attack may later lead to legal difficulties. Operational hesitation is unacceptable.

It is believed others have been taken out in a more low-key fashion. This policy on assassination will be expanded as more U.S. citizens are attacked and killed. The expansion will occur regardless of who is the President. Opposition to such extraordinary measures, now vocal, will dwindle over time. It will be the actions of the terrorists that necessitate more extreme measures. Most important, the terrorists hold their fate in their own hands. Cessation of acts of terrorism would bring about an end to the need for assassination. It's as simple as that.

Far more complex is the problem of reciprocity. While many Americans have no philosophical problem with our agents eliminating terrorists in other countries, the mere thought of foreign assassins eliminating terrorists in the United States is an anathema. If obtaining international agreement on open use of force against tyrants is difficult, establishing mutually acceptable rules of engagement for assassination will be nearly impossible.

THE NON-LETHAL CHEMICAL PARADOX ADDRESSED

The non-lethal weapons paradox has been brought to the attention of Congress. On February 5, 2003, Secretary of Defense Rumsfeld addressed the problems imposed on use of non-lethal chemical systems by the 1993 Chemical Weapons Convention. In his presentation, Secretary Rumsfeld noted that it would be beneficial to be able to use various non-lethal chemicals for operations such as clearing caves. The use of NLWs, he pointed out, could save lives of both our troops and those hidden in the caves.

Rumsfeld also indicated there would be similar problems during urban operations. Given concerns about holding down casualties should fighting occur in Baghdad, the use of these chemicals would allow our forces to clear buildings with minimal risk to noncombatants and themselves.

The Russian use of a chemical agent they called M99 was briefly inserted into the manuscript after it was first employed on October 27, 2002. While it proved effective in incapacitating everyone within a few seconds, approximately 120 people died. As a result of those deaths, the press generally vilified the agent and the people who used it. The headlines were wrong. They should have read, "Six Hundred Hostages Freed." Two steps could have been taken that would have reduced the number of fatalities. The first would have been to notify a doctor at medical facilities that they might receive a large number of patients and that they should be treated like heroin overdoses. That action would have preserved the identity of the specific agent, while providing physicians with sufficient information to initiate adequate treatment.

The second step would have been to have hundreds of caregivers standing by to administer CPR or other resuscitation procedures. It was known that fentanyl, the primary ingredient of M99, is a respiratory inhibitor. Once the Spetsnaz troops had secured the site, other people could have been assigned to ensure they kept the victims breathing until they reached medical facilities. As seen on television, unconscious victims were placed on buses without any attention being paid to them.

Our report from the National Research Council on non-lethal weapons and technology, which was published January 2003, addressed calmative agents. The report noted that U.S. research on these agents had been stopped long ago due to concerns about chemical treaty violations.

Development of calmative chemical agents is a critical topic that must be addressed. The treaties must be changed. Failure to do so will result in the deaths of many people. Those who wish to adhere to strict interpretation of outdated thinking should be held accountable for their actions. As mentioned earlier, it seems ironic that agencies associated with peace movements are those most opposed to logical and life conserving changes to chemical treaties. They leave the use of overwhelming lethal force as the only viable alternative. The *slippery slope* argument fails the test of logic. If we can save lives through the use of advanced chemical or biological agents, we should be bound morally to do so.

Some recent articles have suggested that non-lethal weapons such as these lead to *fuzzy ethics*. The reality is this thinking is *fuzzy logic*. The notion that all chemical or biological technologies are inherently bad if applied in conflict is both unrealistic and wrongheaded. Any all-encompassing condemnation of a technology sector lacks common sense and invariably produces unintended consequences. As noted in the foregoing discussion, conservation of life with non-lethal weapons is banned by law. Again, intent of the human user should be the fundamental issue.

ENERGY REVISITED

If there is good news that has come about in recent months, it is the acknowledgment that alternative energy is now seen as high priority. We have seen how a single disruption in oil supplies can affect the economy. Finally, we have heard senior officials commenting on new sources of fuels. Some of these are outlined previously. We know how to produce them and should get on with it.

However, we have stopped well short of a comprehensive program searching for breakthrough technologies. The promise is there. Frankly, our national survival probably depends on a dramatic shift away from dependence on fossil fuels. There is little doubt that a concerted scientific effort would break our reliance on petrochemicals. With that, we would bring about new relationships in the global community. Energy independence is the asymmetric strategy that wins the war.

OPERATION IRAQI FREEDOM

Only the timing was a bit unexpected when an artfully orchestrated precision attack rained bombs and cruise missiles on a single building in Baghdad. Enticing intelligence reports placed Saddam at that building. With decapitation too promising to resist, the hastily organized raid heralded the onset of the much anticipated action to remove his oppressive regime from power. Two days later, U.S. Marines and Army armored divisions supported by British units roared out of Kuwait and began the most successful invasion in the history of warfare. Bypassing cities that contained Iraqi forces, the spearheads dashed in record time for nearly 300 miles to the doorstep of the capital city. On April 9, 2003, it fell to the coalition forces and Iraqi people, and the symbols of power were publicly destroyed as the faux-elite Republican Guard units spontaneously melted away.

Though occurring late in the editing process of this book, the impact of these events cannot go unnoticed. In fact, these actions have borne out some predictions and served as a harbinger for others. They do not change the basic premises. Extremely well planned and executed, Operation Iraqi Freedom demonstrated the utility of heavy armor and relied on both prepositioned material and long-duration transshipment of more units. While Special Operations forces had been active for some time, conventional units took months to deploy.

The political wrangling preceding this operation demonstrated conclusively the need for the capability to attack unilaterally without massive international cooperation and concurrence. Machinations of the Turkish Parliament prevented much needed armor from being deployed in the north, which inevitably prolonged the conflict and placed light forces there at risk. King Abdullah of Jordan schizophrenically allowed limited engagements from his soil while simultaneously bowing to domestic discontent and denouncing the actions.

The news media, both foreign and domestic, reported on the war with decidedly political bias. Outlets such as *al Jazeera* in the Middle East carried stories that were knowingly false and graphically incongruent. Until the fall of Baghdad, they reported the preposterous claims of the notorious Iraqi Information Minister, who became known as *Baghdad Bob*. With tanks of the 3rd Mechanized Infantry Division parked only a block away, he claimed, and the Arab press reported, that no Americans were in the city and our forces were being killed in large numbers. Pictures of injured children accompanied nearly every story they ran. Even press in the U.S. and Briton ran stories that were spun to fit preestablished editorial policies.

In a brilliant stroke, the Department of Defense developed a process of embedding reporters with troop units. This has done much to raise awareness of reporters and the public about the nature of combat operations. The relatively short initial liberation phase provided the world with the first around-the-clock television coverage. It also demonstrated that war when viewed through a narrow focus might miss the bigger picture. Still, it was probably the best thing that has happened to the media regarding conflict reporting. The question that remains is whether we are prepared to watch soldiers die in real time. I suspect not.

The strategic impact of Operation Iraqi Freedom will continue for decades. It is most likely that relations between the people of the Middle East and the United States and United Kingdom will fester for decades. That will be true even though massive aid will be given to the area and most of the poeple of Iraq wished to be rid of Saddam's corrupt regime. Other international relations, especially with European companies that illegally traded with Iraq during the embargo, will remain tenuous. Certainly the role of the United Nations in deterring terrorism and aggression has been significantly diminished.

Receiving only minimal attention in the media was the grenade attack by Sergeant Asan Akbar on a brigade command post of the 101st Airborne Division in Kuwait. The significance should not go unnoticed. The attack was not the actions of a single deranged individual. Rather, it was the first of similar acts that are likely to follow. *Future War* predicted that people belonging to multiple organizations—and everyone does—would encounter competing, and sometimes incompatible, belief systems. In coming conflicts, the military will be faced with the possibility of questioning the allegiance of soldiers. While there have always been cases of individuals committing criminal acts, this is different. We are approaching a time at which loyalty must be assured, not assumed. The acts of terrorism will not be isolated. It is likely that we will experience other attacks, such as suicide bombings, perpetrated by Americans within our country.

Winning the peace will be far harder than winning the war. With tensions between the Middle East and America exacerbated, the need for energy independence ascends in priority. Calls for religious and social harmony, however noble, are swimming against history. Complex democratic forms of governments cannot be bestowed on people ill-prepared to exercise them. The root issue is a fundamental conflict of values. We Americans are a magnanimous poeple who would prefer to be loved rather than feared. Unfortunately, there are others who respect only those they fear. So be it.

APPENDIX:
THE IMPORTANCE OF DEFINING THE CONFLICT

ARE WE LOSING THE strategic conflict? Following the attacks of September 11, President Bush addressed the U.S. Congress to announce *a war on terrorism*. That war would engage the perpetrators and those who supported them wherever they may be. While the terrorists had struck at targets in the United States, the President was defining the conflict in broader terms. In so doing he hoped to both broaden the scope of the conflict while establishing a focus on specific individuals and deeds. The intent was to attempt to identify and isolate the terrorists from the Muslim world. Since then Western leaders have repeatedly stated that we are not at war with Islam.

Osama bin Laden and al Qaeda countered by calling for *jihad* and worked diligently to unite all those of the Islamic faith and to bring them into the conflict. What they are doing, very effectively, is redefining the conflict for their strategic advantage. Their goal appears to be to make this a global holy war, one with devastating consequences for Western society.

However well intended, U.S. efforts such as dropping food to Afghan refugees are having little, if any, effect on the Muslim people around the world. Highly restricting targets to limit collateral casualties has not been believed. There have been many statements of support for the U.S. position by heads of state from Arab and Muslim countries. These official condemnations of terrorism are directly contradicted by popular demonstrations either in support of Osama bin Laden or denouncing the actions of the United States.

A fundamental problem in many Muslim-dominated countries is that the current governments are viewed internally as corrupt and kowtowing to unacceptable Western values. In short, large segments of these populations believe their leaders have sold out Islam, often for petrodollars.

This is not an isolated phenomenon. Countries from Algeria to the Philippines are experiencing varying degrees of domestic strife. Bin Laden is reportedly assuming cult status in Africa. Small anti-U.S. demonstrations were held in Indonesia, the fourth most populous country in the world, but one already experiencing enormous discord.

Pakistan is experiencing daily riots by bin Laden supporters. Although the president, General Musharraf, is cracking down, it should be remembered that he came to power in a coup and has only limited popular support. It is also noteworthy that Pakistan's economy is weak and that they have nuclear weapons.

More important may be one of bin Laden's key objectives: bringing down the House of Saud and, with it, cutting off vast oil supplies so essential to Western nations. The Saudi Arabia government has had little popular support for many years. The government's internal strategy has been to provide people with sufficient material goods and services so that they will remain complacent. In fact, they have traditionally imported soldiers from Pakistan for their defense needs. The rulers feel that it is better to have foreign Muslim troops under arms than trust training their own citizens, who might then attempt a coup.

Americans tend to think that having our troops in Saudi Arabia adds stability to the region and provides defense against aggressors such as Iraq. Our service members serve at both financial cost to the United States and personal sacrifice. However, many Muslims believe that the mere presence of those troops in the country of Mecca and Medina is an insult to Islam, and a stated objective of al Qaeda is to remove our troops from the area.

Throughout the Muslim world there are many disenfranchised young men who are willing to join this cause and sacrifice their lives if necessary. Many Muslims who do not support terrorism consider Western hegemony to be a greater threat. They are aware that for many years we historically have supported many of the leaders they believe to be corrupt and repressive. In their opinion, we have blindly sided with Israel, at their expense.

If bin Laden is successful in his bid to redefine this conflict as a *jihad*, the enormity of the implications is almost unimaginable. There are no simple solutions. To regain the initiative, Western leadership must form a strategy that defines the conflict in a manner that not only appeals to governments of Muslin-dominated countries but also is embraced by the general population. Only then will we be able to separate al Qaeda from the vast support base that is potentially theirs. That strategy must incorporate large social organizations based on belief systems and not rely on traditional nation-state relationships. Failure to do so will allow unacceptable expansion of the war.

Who defines this conflict wins.

John B. Alexander
Published in the *Las Vegas Sun*
October 18, 2001

NOTES

1. John A. Warden III, Col. USAF (Ret.), *The Air Campaign*: *Planning for Combat* (National Defense University Press, 1988).
2. Mark Bowden, *Black Hawk Down* (Penguin, 1999).
3. Pres. George W. Bush, "Address to a Joint Session of Congress and the American People," September 20, 2001.
4. Bob Woodward, *Bush at War* (Simon and Schuster, 2002).
5. Former Secretary of Defense Caspar Weinberger, *Larry King Live,* transcript of interview, December 5, 2001.

_____ 1. "PHASERS ON STUN"

1. John Barry's "Soon, Phasers on Stun," (*Newsweek,* February 7, 1994) was probably the first of many such articles.
2. Edward Teller recounted this in private conversations with author at Los Alamos National Laboratory in 1993.
3. John B. Alexander, *Future War: Non-Lethal Weapons in Twenty-first-Century Warfare* (St. Martin's Press, 1999).
4. Complete details about the ABL programs from official sources can be found at www.airbornelaser.com.
5. www.airbornelaser.com.
6. Jim Riker, *OSDR-99-20: Airborne Tactical Laser (ATL)—Performance Details,* Office of the Secretary of Defense, October 15, 1999.
7. Miguel Navrot, "Kirtland Developing a Gunship," *Albuquerque Journal,* March 5, 2002.
8. Associated Press, "Army's High-Speed Laser Hits Shell," November 9, 2002.
9. John Chipman, "U.S. Energy Beam Lightly Scorches Protestors: Easier than an A-bomb," *National Post,* March 2, 2001.
10. Christopher Castelli, "Questions Linger About Health Effects of DOD's Non-Lethal Ray," *Inside the Navy,* March 26, 2001.

11. John Yauckey, "Crowd Control Cookery: Microwaves Among New Non-Lethal Weapons," Gannett News Service, April 1, 2001.

12. Col. George Fenton has participated in many non-lethal weapon conferences. The pulsed energy projectile is part of his standard presentation of emerging technologies.

13. Craig Jensen, private briefings at Mission Research Corporation facilities in Los Alamos, New Mexico, February 22, 2001.

14. *Collateral Damage,* Warner Brothers Studios.

15. "Topic: In-custody Deaths," *TASER International News Bulletin,* February 2002.

16. *National Vital Statistics Reports* 49, no. 8, http://www.cdc.gov/nchs/fastats/druguse.htm.

17. Theodore Chan, M.D., et al., *Pepper Spray's Effects on a Suspect's Ability to Breathe,* National Institute of Justice report, November 2001.

18. Ed Vasel, PepperBall Tactical Systems, private conversation, May 22, 2002.

19. John Alexander, "The Taser Alternative," *Washington Post,* October 1, 1999.

20. Colin Burrows, "Operationalizing Non-lethality: A Northern Ireland Perspective," *Journal of Medicine, Conflict, and Survival* 17, no. 3 (July–September 2001).

21. Jay Kehoe of LE Technologies provided these data. He is also a lieutenant in the Glastonbury, Connecticut, police department.

2. YOU CAN RUN, BUT YOU CAN'T HIDE

1. *Clear and Present Danger,* Paramount, 1994.

2. *Behind Enemy Lines,* TM and 2001 Fox, 2001.

3. *Enemy of the State,* Touchstone, 1998.

4. Stephen P. Aubin, "The 'Cover-Up' Story That Made News All Over. The Problem Was That It Wasn't so," *Air Force Magazine,* July 2000, Vol. 83. No 7.

5. *Newsweek,* May 15, 2000.

6. David Ruppe, "Lessons Learned from the Kosovo War," *Newsweek,* April 14, 2000.

7. Robert Roy Britt, "Satellites Play Crucial Roles in Air and Ground Battles," Space.com, October 9, 2001.

8. "DSP mounted on IUS, MILNET: Satellites—Defense Support Program (DSP)," http://www.milnet.com/milnet/dsp/htm.

9. Glenn W. Goodman, Jr., "Assured Access to Space," *Armed Forces Journal,* July 2002.

10. "Space-Based Infrared—Low Space and Missile Tracking System, Brilliant Eyes," http://www.fas.org.

11. Emily Clark, "Military Reconnaissance Satellites (IMINT), the Center for Defense Information," http://www.cdi.org/terrorism/satellites-pr.cfm.

12. Dave Monitz, "Winter's Cold May Help Military Track Taliban," *USA Today,* November 4, 2001.

13. Britt, "Satellites Play Crucial Roles."

14. Clark, "Military Reconnaissance Satellites (IMINT)." A good explanation of how AR and Doppler-shifted radar works can be found at the Federation of American Scientists Web site, http://www.fas.org/spp/military/program/imint/lacrosse.htm.

15. Federation of American Scientists, "Improved Crystal," http://www.fas.org.

16. Space and Technology, "Atlas Successfully Launches Classified NRO Satellite," October 11, 2001, http://www.spaceandtech.com/digest/flash2001/flash2001-090; shtml.

17. "Lacrosse," http://www.collections.ic.gc.ca/satellites/english/function/reconnai/lacrosse.htm.

18. Clark, "Military Reconnaissance Satellites (IMINT)."

19. "Unmanned Aerial Vehicles," *DTIC Review* AD-A351447.

20. "RQ-1 Predator Unmanned Aerial Vehicles," U.S. Air Force Fact Sheet.

21. Federation of American Scientists, "RQ-1 Predator MAE UAV," http://www.fas.org/irp/program/collect/predator.htm.

22. "Report Details SEAL's Last Stand in Afghanistan," CNN, May 17, 2002.

23. Victoria Samson, "Q&A on the Use of Predator in Operation Enduring Freedom," Center for Defense Information, February 11, 2002.

24. "Global Hawk," U.S. Air Force Fact Sheet.

25. "Operational Debut for Global Hawk," *Jane's Yearbook* entry, October 8, 2001.

26. Northrop-Grumman, "Global Hawk Data Sheet," http://www.is.northupgrumman.com/products/usaf products/global hawk/global hawk.html.

27. Raytheon, "Global Hawk Integrated Sensor Suite," http://www.raytheon.com/es/esproducts/sesghk/sesghk/htm.

28. Ivan Amato, "Future Tech: Beyond X-ray Vision, Can Big Brother See Right Through Your Clothes?" *Discover,* 23, no. 7 (July 2002).

29. "About Millimeter Waves," www.millivision.com.

30. "A Camera and Algorithm Know It's You," *MIT Technology Review,* November 2001.

31. FERET Overview, Department of Defense Counterdrug Technology Program Office, November 2001.

32. "U.S. Pours Money into Face Recognition," *Reuters,* April 12, 2002.

33. John Schwartz, "New Side to Face-Recognition Technology: Identifying Victims," *New York Times,* January 15, 2002.

34. Robert Parks, "Liars Never Break a Sweat," *New York Times,* July 12, 1999.

35. Ibid.

36. Mark S. Zaid, "Failure of the Polygraph," *Washington Post,* April 16, 2002.

37. Ibid.

38. Cleve Backster, private conversations over twenty years.

39. Alan P. Zelicoff, "Polygraphs and National Labs: Dangerous Ruse Undermines National Security," *Skeptical Inquirer,* July/August 2001. Daniel King is the correct name, not David King as listed in the article.

40. Capt. Charles "Sid" Heal, multiple private conversations.

41. Margie Wylie, "Police Use of Voice Stress Analysis Generates Controversy," Newhouse News Service, November 6, 2001.

42. Harry Rosen, *LEADS Business Plan,* July 2002.

43. John Alexander, personal observation of the taped television interview and LEADS analysis.

44. Julia Sheeres, "Thought Police Peek into Brains," *Wired,* October 5, 2001.

45. Mary Vallis, "Brain Scan Can Detect Lies, Researchers Find," *National Post,* November 12, 2001.

3. THE LETHAL LEGACY

1. Richard Norton-Taylor, "Taliban Hit by Bombs Used in Vietnam," *Guardian,* November 7, 2001.

2. "BLU-118/B—Thermobaric Warhead," Defense Threat Reduction Agency Fact Sheet.

3. Federation of American Scientists, "BLU-82B," www.fas.org/man/dod-101/sys/dumb/blu-82.htm.

4. "Pentagon to Use New Bomb on Afghan Caves," CNN.com, December 23, 2001.

5. Victorino Matus, "During Operation Anaconda, the United States Rolled Out a New Weapon—the Thermobaric Bomb. It's Worse than a Daisy

Cutter and It May Have Saddam's Name on It," *The Weekly Standard*, March 12, 2002.

6. Vernon Loeb, "Concrete-Piercing Bombs Hammer Caves," *Washington Post*, December 13, 2001.

7. 1st Lt. Kimberley Devereux, "B-1B Drops First GPS-Guided JDAM," Public Affairs, Wright-Patterson Air Force Base, March 1998.

8. Paul Eng, "Guided by the Stars U.S. Uses New Precision Bomb Guided by Satellites," ABCNEWS.com, October 9, 2001.

9. Staff Sergeant Bob Pullen, "Andersen Makes 'Dumb' Bombs 'Smarter' with New Guidance Kits," Public Affairs, Andersen Air Force Base, September 7, 2000.

10. "Tragic Turn Pentagon: Three U.S. Soldiers Killed in Friendly Fire Incident," *ABCNEWS.com*, December 5, 2002.

11. Robert Roy Britt, "Satellite-Guided Bomb Misses Target, Kills 4 Afghan Civilians," *Space.com*, October 14, 2001.

12. John Lumpkin, "Pentagon Proposes Power Boost in Future GPS Navigation Satellites," *Associated Press*, May 7, 2002.

13. Lloyd "Bud" Kinsey, Raytheon, Tucson, Arizona in several briefings and private discussions with the author about the Big Gun concept development.

14. Jonathan Weisman, "Nuclear Arsenal Upgrade Planned 'Bunker Buster' Marks a Shift in U.S. Strategy," *USA Today*, March 19, 2002.

15. Christopher Bolkcom, *Army Aviation: The RAH-66 Comanche Helicopter Issue, Report for Congress*, Congressional Research Service, January 9, 2002.

16. Secretary of the Navy John Dalton, Chief of Naval Operations Admiral J. M. Boorda, and Commandant, U.S. Marine Corps, General Carl E. Mundy, *Forward . . . from the Sea*, 1992.

17. Those politically correct will suggest that the Saudi princes did everything we asked of them in Enduring Freedom. However, the U.S. State Department was careful not to officially ask for takeoff rights or permission to use the command and control center built by U.S. Central Command because they knew it would be refused.

18. Status of U.S. Navy as of August 12, 2002, *Navy Fact File*.

19. Marty Kauchak, "Navigating Changing Seas," *Armed forces Journal International*.

20. "August 2002 John Pike, Streetfighter/Sea Lance," http://198.65.138.161/military/systems/ship/streetfighter.htm.

21. "Arleigh Burke Class Destroyers," *Naval Institute Press*, http://www.usni.org/arleighburkeddg.htm.

22. Harold Kennedy, "Army Overhauls Its 70-Ton Behemoth—the Abrams Tank," *National Defense*, September 2001.

23. "Crusader, 155-Self Propelled Howitzer Program," www.Army Technology.com/projects/crusader.

24. I participated as a subcontractor in the proposal development and submission for both the FCS and OFW.

25. "Stryker Family of Vehicles Fact Sheet," http:/www.army.mil/features/stryker/stryker_spec.pdf.

26. Dutch Degay, Special Briefing on Objective Force Warrior sponsored by Army Public Affairs, May 23, 2002. (Degay is the Natick Laboratory OFW program manager.)

27. Brad Lemley, "Really Special Forces: A Powered Exoskeleton Could Transform the Average Joe into a Supersoldier," *Discover* 23, no. 2, February 2002.

28. Andrew Bridges, "First Flight of X-45A Robot Combat Plane Called Success," Associated Press, May 24, 2002.

29. Sandy Riebling, "Unmanned Aerial Vehicles," *Redstone Rocket* 51, no. 28 (July 17, 2002).

30. Department of Defense, "Advanced Concept Technology Demonstration List for 2002 Announcement," March 5, 2002.

31. James M. McMichael and Col. Michael Francis, USAF, "Micro Air Vehicles—Toward a New Dimension in Flight," www.darpa.mil/tto/MAV/mav auvsi.html.

32. Barbara Fletcher, "UUV Master Plan: A Vision for the Navy UUV Development," Space and Naval Warfare Systems Center, 2000.

33. Robert I. Wernli, "Recent U.S. Navy Underwater Vehicle Projects," Space and Naval Warfare Systems Center, 2002.

34. Roxana Tiron, "High-Speed Unmanned Craft Eyed for Surveillance Role," *National Defense*, May 2002.

35. "Robots Doing Dirty Work for the Army in Afghanistan," Yahoo News, July 30, 2002; and Nic Robinson, "Meet Packbot: The Newest Recruit," CNN News, August 1, 2002.

36. "Department of Defense Joint Robotics Program," http://www.jointrobotics.com/webdoc/brochure.

37. Information about Sandia robotic vehicles can be found on their Web site: http://www.sndia.gov/isrc/Capabilities/Integration Technologies.

38. Adm. David Jeremiah, U.S. Navy (Ret.), "Nanotechnology and Global Security," paper presented at the Fourth Foresight Conference on Molecular Nanotechnology, November 9, 1995.

39. John L. Petersen and Dennis M. Egan, "Small Security: Nanotechnology and Future Defense," *Defense Horizons,* National Defense University, March 2002.

40. Chloe Veltman, "Nanotech Future for Soldiers," BBC News, September 21, 2001.

41. Phillip Anton et. al., "The Global Technology Revolution," RAND, 2001.

42. Ralph C. Merkle, "Nanotechnology Is Coming," http://www.merkle.com/papers/FAZ000911.html.

43. W. M. Tolles, "Self-Assembled Materials," *Materials Research Society Bulletin* 25, no. 10 (October 2000).

44. Ibid.

45. Andrew Chen, "The Ethics of Nanotechnology," http://www.scu.edu/Ethics/publications/submitted/chen/nanothechnology.html.

46. Erik Baard, "Microscopic Nanotubes Could Make Ships lightweight, Superstrong," *SPACE.com,* February 6, 2002.

47. Peter Kupfer, "Ultimate Spy—Airborne Smart Dust Sees, Hears, . . . & Reports," *San Francisco Chronicle*, November 20, 2000.

48. "Smaller, Lighter, Cheaper," *Jane's International Defense Review,* May 2001.

4. THE NILE

1. CNN Interactive, "Security Tightened in Egypt After Massacre," www.CNN.com/WORLD/97/11/18/egypt.attack.mubarack, November 18, 1997.

2. Sameh Taha was our tour guide in Egypt. He was an invaluable source of information. Most of the statistics and comments about the state of society in Egypt come from hours of conversation with this most remarkable man. Unlike many guides, he was prepared to show the good, the bad, and the ugly. Like many others in the tourism industry, he was severely hurt by the events of 9/11.

3. John J. Bentley, *Egypt Guide* (Open Road, 1988).

4. M99 was the incapacitating agent first employed by the Russians on October 27, 2002. It is a fentanyl derivative with other chemicals mixed together for rapid incapacitation. However, opiates are respiratory inhibitors and require immediate and constant attention so that victims do not cease breathing. All of the victims who made it to medical care survived the Moscow raid.

5. PORTS OF CALL

1. Steve Vogel, "I Just Didn't Want to Die on the Ship," *Washington Post*, October 28, 2000.
2. Roberto Suro and Thomas E. Ricks, "Terrorism Suspected in Navy Ship Attack," *Washington Post*, October 13, 2000.
3. Department of Defense, "USS *Cole* Commission Report," http:/www.defenselink.mil/pubs/cole20010109.html.

6. ANOTHER WORLD

1. Claire Marshal, "Peru Set to Be Drug Leader," BBC News, February 17, 2001.
2. Rob Rachowiecki, *Peru: A Lonely Planet Travel Survival Kit*, Lonely Planet, Hawthorne, Australia, 1996.
3. Howard Lawler, Ph.D., personal conversations in Amazon area of Peru, January 2001.
4. Tonya Sissel, "Michigan Missionaries Killed in Peru," *Compass* 50, no 27 (April 27, 2001).
5. CNN, "Plane Shootdown: Drug Intercept Flights Suspended in Peru," www.cnn.com/2001/US/04/21/peru.plane.02/.
6. Drug Enforcement Agency, Intelligence Division, Strategic Intelligence Section, *Coca Cultivation and Cocaine Processing: An Overview*, September 1993.
7. James Read, "Fujimori's Controversial Career," BBC News, September 18, 2000.
8. Lt. Col. Stephen P. Howard, USAF, "The War on Drugs: Two More Casualties," *Aerospace Power Journal*, Winter 2001.

9. THE WAR WE WILL FIGHT

1. Colonel John A. Warden III, *The Air Campaign: Planning for Combat* (National Defense University Press, 1988).
2. General Norman Schwarzkopf, *It Doesn't Take a Hero*, (Bantam, 1993). In this book, General Schwarzkopf gives credit to Colonel Warden for the air campaign planning for the Gulf War.
3. Hal Puthoff, Ph.D., "Ground State of Hydrogen as a Zero-Point-

Fluctuation-Determined State," *Physical Review D:* Particles and Fields, May 15, 1987.

4. Hal Puthoff, Ph.D., "The Sea of Quantum Energy in Which We Live: The 'Holy Grail' of 21st Century Science and Technology," Institute for Advanced Studies at Austin.

5. Colonel Gary Anderson, USMC (Ret.), "Urban Warrior," in *The City's Many Faces,* edited by Russell Glenn (RAND, 1999).

6. Convention on the Prohibition of the Development, Production, Stockpiling and Use of Chemical Weapons and on Their Destruction, January 13, 1993.

7. Nicholas Warr, *Phase Line Green: The Battle for Hue,* (1968; reprint, Naval Institute Press, 1997).

PART III. PLAN A: WIN THE WAR ON TERROR

1. John B. Alexander, *Future War: Non-Lethal Weapons in Twenty-first-Century Warfare* (St. Martin's Press, 1999).

2. Secretary of Defense Donald Rumsfeld, "Transforming the Military's Foreign Affairs" *Foreign Policy,* May–June 2002.

10. THE ROOT OF ALL EVIL

1. Jonathan Wells, Jack Meyers, and Maggie Mulhill, "Saudi Elite Linked to bin Laden Financial Empire," *Boston Herald,* October 14, 2001.

2. Gen. Barry McCaffrey, director of ONDCP, *The Drug Legalization Movement in America: Testimony Before the House Government Reform and Oversight Committee, Subcommittee on Criminal Justice, Drug Policy, and Human Resources,* June 16, 1999.

3. National Drug Control Strategy, Budget Summary 2000, http://www.ncjrs.org/ondcppubs/publications/policy/budget00/exec_summ.html.

4. "Killing Pablo," *Philadelphia Enquirer,* November 20, 2001.

5. David R. Henderson, Ph.D., "Ending the U.S. Role in Colombia's Drug War Would Weaken Terrorism and Strengthen Democracy," *San Francisco Chronicle,* November 7, 2001.

6. Timothy Lynch, "After Prohibition: An Adult Approach to Drug Policies in the 21st Century," CATO Institute.

7. Ludwig von Mises Institute, "Making Economic Sense," www.mise.org/econsense/ch91.asp.

8. Carol Martin, "Diamond Rush," CBSNEWS.com, August 21, 2002.

9. A. N. Lecheminant et al, "Diamonds! 'The Great Canadian Diamond Rush North of Yellowknife,' " *McNellis,* 1993.

10. Ambassador David C. Newsom, "Diamonds Fuel African Conflict," *Christian Science Monitor,* July 14, 1999.

11. Ibid.

12. Jake Sherman, "Blood Diamonds Fund More than Terrorism," *Washington Post,* November 7, 2001.

13. Edward Jay Epstein, "Have You Tried to Sell a Diamond?" *Atlantic Monthly,* 249, no. 2, (February 1982).

14. Jani Roberts's *Glitter and Greed: The Diamond Investigation* was based on research for a documentary for *Frontline* that was never aired.

15. Congressman Tony Hall (Democrat Ohio), appearing on "Dirty Diamonds," *Background Briefing,* produced by Stan Correy, Australia, February 20, 2000.

16. World Diamond Council International Headquarters Web site http//: www.worlddiamondcouncil.com.

17. "New Rules Target 'Blood Diamonds,' " CBSNEWS.com, Antwerp, Belgium, July 19, 2000.

18. Ibid.

19. Margarita Antidz, "Bloodied Diamonds to Become Scarcer," *Reuters,* Moscow, June 13, 2001.

20. Ian Smilie, appearing on "Dirty Diamonds," *Background Briefing,* produced by Stan Correy, Australia, February 20, 2000.

21. Billy Steele, "Money Laundering: A Brief History," http:// www.laundryman.u-net.

22. John McDowell and Gary Novis, "The Consequences of Money Laundering and Financial Crime," *Economic Perspectives,* May 2001.

23. Ibid.

24. Nigel Morris-Cotterill, "Think Again: Money Laundering," *Foreign Policy,* May/June 2001.

25. J. McDowell and Novis, "The Consequences of Money Laundering and Financial Crime."

26. Morris-Cotterill, "Think Again."

27. Ibid.

———— **11. POWER OF THE PRESS—A STRATEGIC WEAPON**

1. Ali Akbar Dareini, "Iran Says It Won't Stand Idle If Iraq Is Attacked," Associated Press, September 2, 2002.

2. "Russia Warns of Veto on Iraq," BBC, http://news.bbc.uk/1/hi/world/middle_east/2231483.stm, September 2, 2002.

3. Mike Cohen, "Mandela Blasts U.S. Attack Threats," Associated Press, September 2, 2002.

4. Saddam Hussein, President of Iraq, in private E-mail to Christopher Love in Pennsylvania, October 18, 2001. This E-mail was widely circulated and received attention from most major news outlets. I did track it back to the official Iraqi News Service and believe the contents to at least have been approved by Saddam Hussein.

5. Juliet Eilperin, "Democrat Implies Sept. 11 Administration Plot," *The Washington Post,* April 12, 2002.

6. Chuck de Caro, "Avenging 9/11: Defeating a New Kind of Enemy," AEROBUREAU, 2002.

7. *Simone,* starring Al Pacino, *New Line Cinema,* August 2002.

8. Brian Ross, "Journalists Smuggle Depleted Uranium into New York," ABC News, September 6, 2002.

9. "BA Security Lapses Exposed," Reuters, September 18, 2002.

10. Margaret and Melvin DeFleur, "The Next Generation's Image of Americans; Attitudes and Beliefs Held by Teen-agers in Twelve Countries," Boston University, September 2002.

11. John Alexander, "Optional Lethality: Evolving Attitudes Towards Non-Lethal Weapons," *Harvard International Review,* Summer 2000, is where this term was introduced.

12. "Is It safe," interview with Norman Mineta on *60 Minutes II,* October 24, 2001.

13. "The Safest Airline," CBSNEWS.com, August 21, 2002.

14. David T. Zabecki, "The Vietnam War Film *We Were Soldiers* Has Been Slammed by Some Reviewers as a Revisionist Movie. It Is." *Vietnam Magazine,* October 2002.

15. Seymour Hersh, "What Happened in the Final Days of the Gulf War?" *New Yorker,* May 22, 2002.

16. "Hurricane Carter: The Other Side of the Story," http://graphicwitness.com/carter/.

12. THE EPITOME OF PRECISION

1. Col. John Warden, USAF (Ret.), private conversations.
2. Eric Haney in private E-mails and conversations. His book, *Inside Delta Force: The Story of America's Elite Counterterrorism Unit* (Delacorte Press, 2002), provides excellent reading for an understanding of how the United States has responded to terrorism.
3. David Gow, "Bush Gives Green Light to CIA for Assassination of Named Terrorists 'Covert Killings to Take In Less Important al-Qaida figures,' " *Guardian,* October 29, 2001.
4. Ralph Peters, *Beyond Terror* (Stackpole, 2002).
5. Fox News, "Ashcroft Faces Objections to Death Penalty in Europe," December 12, 2001.
6. Molly Moore and John Anderson, "Suicide Bombers Give an Edge to Palestinians," *The Washington Post,* August 19, 2002.
7. Holger Jensen, "A Long History of Terrorism," Scripps Howard News Service, September 29, 2001.
8. Thomas B. Hunter, "Wrath of God: The Israeli Response to the 1972 Munich Olympic Massacre," *Journal of Counterterrorism & Security International* 7, no. 4 (Summer 2001).

PART IV. PLAN B: THE EVENT HORIZON

1. "Understanding Islam and the Muslims," Islamic Affairs Department, Embassy of Saudi Arabia, Washington, D.C.
2. Shahid Athar, M.D., "Twenty-five Questions About Islam," http://www.islam/usa/25ques.html.
3. Dr. Paulo Paniago, "Terrorism: Trends and Perspectives," IACP Executive Policing Conference, Brasilia, August 27, 2002.
4. Rachel Zoll, "Survey Finds Rise in Catholics, Drop in Liberal Protestant Congregation," Associated Press, September 18, 2002.
5. Anayat Durrani, "The Search for Truth," April 2, 1999.
6. Athar, "Twenty-five Questions About Islam."
7. Samuel P. Huntington, "The Clash of Civilizations?" *Foreign Affairs,* Summer 1993.
8. Andrew Sullivan, "This Is a Religious War," *New York Times,* October 7, 2002.

9. Associated Press, *"Pope Speaks of "Clash of Civilizations,"* Vatican City, November 29, 2002.

10. Thomas L. Friedman, *The Lexus and the Olive Tree* (*Farrar*, Straus, and Giroux, 1999).

11. Robert D. Kaplan, *Warrior Politics: Why Leadership Demands a Pagan Ethos* (Random House, 2002).

12. John B. Alexander, *Future War: Non-Lethal Weapons in Twenty-first-Century Warfare* (St. Martin's Press, 1999).

13. David Wood, "Infantry Losing Money to High-Tech," *Newhouse News Service*, December 8, 2002.

14. Rowan Scarborough, "Fear of Casualties Hampers Hunt for Taliban," *The Washington Times,* December 9, 2002.

13. RETHINKING SPACE MISSIONS

1. Senator Bob Smith, "The Challenge of Space Power," *Airpower Journal,* Spring 1999.

2. Brigadier General Kevin Campbell, meeting with author at Space Command Headquarters, Peterson Air Force Base, October 24, 2000. Discussion indicated that for the next twenty years military space operations would involve satellites and manned space exclusively concerned with the ISS.

3. Maj. Howard Belote, "The Weaponization of Space: It Doesn't Happen in a Vacuum," *Aerospace Power*, Spring 2000.

4. Maj. Gen. (Sel.) John Barry and Col. Darrell L. Herriges, "Aerospace Integration, Not Separation," *Airpower Journal*, Spring 1999.

5. U.S. Space Command, *Vision 2020,* February 1997.

6. Director of Plans, "Control of Space," *US Space Command Long Range Plan,* Peterson Air Force Base, March 1998.

7. Smith, "The Challenge of Space Power."

8. Pres. George W. Bush, "Address to a Joint Session of Congress and the American People," September 20, 2001.

9. Tech Sgt. Lisa Polarek, "Whiteman's B-2s Participate in Air Strikes," *Air Force Link*, October 12, 2001.

10. Joseph Kahn, "Pay Back: U.S. Is Planning an Aid Package for Pakistan Worth Billions," *New York Times*, October 27, 2001.

11. Barry and Herriges, "Aerospace Integration."

12. Lt. Gen. Eugene Tattini, Commander, Space and Missile Systems

Command, private conversation at *Space 2000* conference, Long Beach, California, September 20, 2000.

13. James Oberg, "China Takes Aim at the Space Station," MSNBC.com, October 30, 2001.

14. Frank Oliveri, "Station Cost Management Report Rips ISS Overruns," *Florida Today*, October 17, 2001.

15. NASA Facts, "The TranHab Module: An Inflatable Home in Space," May 1999.

16. George Abbey, "Letter from NASA JCS Center Director: Actions Required to Address ISS Budget Challenges," February 27, 2001.

17. Bob Haltermann, "Bigelow Aerospace Begins Development of Permanent Orbiting Structures," *Space Transportation Association* 3, no. 2 (March 31, 2001).

18. Robert Bigelow, private conversation, December 6, 2001.

19. "Bigelow Aerospace Initiates Preliminary Steps for Private Space Station Launch License," *Space & Technology*, October 2001.

20. "X-33, What is X-33," October 8, 1998, http://stp/msfc.nasa.gov/stpweb/x33/x33about.html.

21. Lt. Col. John E. Ward Jr., "Reusable Launch Vehicles and Space Operations," *Occasional Paper 12*, Center for Strategy and Technology, Air War College, May 2000.

22. Leonard Davis, "NASA Shuts Down X-33, X-34 Programs," *SPACE.com*, March 1, 2001.

23. Marc Millis, "Breakthrough Propulsion Physics Research Program," *NASA Technical Memorandum 107381*, January 26, 1997.

24. *Future Biology Research on the International Space Station* (National Academy Press, 2000).

25. James Beggs et al., "The International Space Station Commercialization (ISSC) Study," Potomac Institute for Policy Studies, Arlington, Virginia, March 20, 1997.

26. Bret Drake, "Human Exploration: A Multi-Destination View," *Space 2000*, September 20, 2000.

27. Leonard David, "Space Tourism in the 21st Century: What Next After Tito," *SPACE.com*, June 29, 2001.

28. Beggs et al., "The International Space Station Commercialization (ISSC) Study."

29. Gen. Thomas S. Moorman Jr., "The Explosion of Commercial Space and the Implications for National Security," *Airpower Journal*, Spring 1999.

30. Lt. Col. Bruce M. Deblois, "Space Sanctuary: A Viable National Strategy" *Airpower Journal* 12, no. 4, (Winter 1998); Lt. Col. Michael Baum, "Defiling the Altar: The Weaponization of Space," *Airpower Journal*, 8, no. 4 (Spring 1994); and Belote, "The Weaponization of Space."

31. Lt. Col. Larry G. Sills, "Space-Based Global Strike: Understanding Strategic and Military Implications," *Occasional Paper 24*, Center for Strategy and Technology, Air War College, August 2001; and Lt. Col. Cynthia A. S. McKinley, "The Guardians of Space: Organizing America's Space Assets for the Twenty-first Century," *Aerospace Power Journal*, Spring 2000.

32. Barry and Herriges, "Aerospace Integration."

33. General Ronald R. Fogleman, "Crossroads in Space," *Airforce Magazine* 84, no. 3 (March 2001).

34. Majors Larry Bell, William Bender, and Michael Carey, "Planetary Asteroid Defense Study: Assessing and Responding to the Natural Space Debris Threat," *ACSC/DR/225/95/4,* Air Command and Staff College, May 1995.

35. Web page on asteroid and comet defense, http://scivis.nps.navy.mil/~swrp/1996Projects/ACDHome.html.

36. Edward Teller, "The Need for Experiments on Comets and Asteroids," in *Proceedings of the Planetary Defense Workshop,* Lawrence Livermore National Laboratory, May 22–25, 1995.

37. Johndale Solem, "Interception and Disruption," in *Proceedings of the Planetary Defense Workshop.* Lawrence Livermore National Laboratory; May 22–25, 1995.

38. Gregory Canavan, U.S. Congressional Hearings on Near-Earth Objects and Planetary Defense, May 21, 1998.

39. Associated Press, "Large Asteroid Passes Close to Earth," January 8, 2002.

40. Col. John A. Warden and Leland Russell, *Winning in Fast Time* (Montgomery, AL: Venturist Publishing, 2002).

41. Albert Harrison, *Spacefaring: The Human Dimension* (Berkeley: University of California Press, 2001).

42. Gen. Michael Ryan and Secretary of the Air Force F. Whitten Peters, *The Aerospace Force: Defending America in the 21st Century,* U.S. Air Force, 2000.

43. National Science and Technology Council, "National Space Policy," The White House, September 19, 1996.

44. Donald Rumsfeld et al., "Report of the Commission to Access United States National Security Space Management and Organization," Pursuant to Public Law 106–65, January 11, 2001.

45. Ibid.

46. Former Secretary of Defense Caspar Weinberger, *Larry King Live,* transcript of interview, December 5, 2001.

47. Lt. Col. Larry J. Schaefer, USAF, *Sustained Space Superiority: A National Strategy for the United States,* Occasional Paper 30, Air University, August 2002. Though this paper was written well after this chapter was drafted and submitted to Aerospace Power, the author came to many of the same conclusions and notes the necessity for U.S. dominance of all aspects of space and to prepare for eventual conflict there.

14. SIX SIGMA SOLUTIONS

1. "Six Sigma, a Division of the American Society for Quality," www.sixsigmaforum.com.

2. Vice Adm. Arthur Cebrowski, "Navy After Next," All Flag Conference, 2000.

3. John B. Alexander, *Future War: Non-Lethal Weapons in Twenty-first-Century Warfare* (St. Martin's Press, 1999). Chapter 20, "Winning," describes in more detail how this important notion will change.

4. Dean Judd, Ph.D., contributed this section on fusion. He is a colleague and retired senior fellow at Los Alamos National Laboratory and former National Intelligence Officer for Science and Technology

5. Harold Puthoff, Ph.D., contributed the work on zero point energy. Working as Director of the Institute for Advanced Physics at Austin, Texas, he is considered one of the foremost authorities in the field.

6. Eric Davis, Ph.D., contributed the material on antigravity. A former associate of mine at NIDS, he has written some of the fundamental material on engineering wormholes for deep space travel.

7. R. L. Forward, "Antigravity," *Proceedings of the Institute of Radio Engineers* 49 (1961): and "Guidelines to Antigravity," *American Journal of Physics* 31, (1963).

8. E. Podkletnov and R. Nieminen, *Physica C* 203 (1992): and E. Podkletnov, "Weak Gravitation Shielding Properties of Composite Bulk $YBa_2Cu_3O_{7-x}$ Superconductor Below 70 K Under EM Field," http://xxx.lanl.gov/abs/cond-mat/9701074 (1997).

9. H. E. Puthoff, "Can the Vacuum be Engineered for Spaceflight Applications? Overview of Theory and Experiments," *Journal of Scientific*

Exploration 12 (1998); H. E. Puthoff, S. R. Little, and M. Ibison, "Engineering the Zero-Point Field and Polarizable Vacuum for Interstellar Flight," *Journal of the British Interplanetary Society* 55 (2002); and E. W. Davis, "Summary of Wormhole/Warp Drive Metric Engineering Research," invited briefings, Lockheed-Martin Company, Houston, TX, and the California Institute for Physics and Astrophysics, Palo Alto, 1999, 2000.

10. E. W. Davis, "Wormhole-Stargates: Engineering the Spacetime Metric," invited briefing, U.S. Air Force Research Lab-Propulsion Directorate, Edwards Air Force Base, California, 2002.

11. Edmund Storms, "Review of the "Cold Fusion" Effect," *Journal of Scientific Exploration* 10, no. 2 (Summer 1996).

12. Ibid.

13. U.S. Department of Energy, "A Nations's Vision of America's Transition to a Hydrogen Economy—to 2030 and Beyond," February 2002, and *Proceedings—National Hydrogen Energy Roadmap Workshop,* April 2–3, 2002.

14. U.S. Department of Energy, "Hydrogen: The Fuel of the Future," http://www.eren.doe.gov/hydrogen/dpfs/hydrofut.pdf.

15. U.S. Department of Energy, "EERE: Hydrogen Energy—Fuel Cells," http://www.eren.doe.gov/hydrogen/dpfs/hydrofut.pdf.

16. Peter Hoffmann, *Tomorrow's Energy: Hydrogen, Fuel Cells, and the Prospects for a Cleaner Planet* (MIT Press 2001).

17. Thomas F. Homer-Dixon, "On The Threshold: Environmental Changes as Causes of Acute Conflict," *International Security,* 16, no. 2 (Fall 1991).

18. Col. Tanzy House, Lt. Col. James Near, Lt. Col. William Shields, Maj. Ronald Celetano, Maj. David Husband, Maj. Ann Mercer, and Maj. James Pugh, *Weather as a Force Multiplier: Owning the Weather by 2025.* This paper was written for the Air Force 2025 project. It does not represent official doctrine.

19. Brig. Gen. Fred Lewis, *Air Force Weather Policy on Weather Modification,* February 13, 1999.

20. Col. House et al., *Weather as a Force Multiplier.*

21. Weather Modification Association, *Weather Modification: Some Facts About Seeding Clouds.* August 1996.

22. Ibid.

23. *Status and Future Directions in U.S. Weather Modification Research and Operations,* National Academy of Sciences Project BASC-U-01-01-A.

24. Trevor James Constable, discussion and observations with the author. Videotapes of these experiments still exist.

25. *Department of State Bulletin 74* 13 (June 1977).

26. John B. Alexander, "The New Mental Battlefield," *Military Review,* December 1980.

27. John B. Alexander, Janet Morris, and Richard Groller, *The Warrior's Edge* (William Morrow, 1990).

28. The International Remote Viewers' Association may be located at http://www.irva.org/.

29. Joseph McMoneagle, *Mind Trek: Exploring Consciousness, Time, and Space Through Remote Viewing* (Hampton Roads Publishing, 1993); *Remote Viewing Secrets: A Handbook* (Hampton Roads Publishing, 2001).

30. Jim Schnabel, *Remote Viewers: The Secret History of America's Psychic Spies* (Dell Publishing, 1997), provides a good overview of the remote-viewing program from an outsider's perspective; Major Paul Smith, U.S. Army (Ret.), *Reading the Enemy's Mind* (Tor, 2003), offers the most complete description of the remote-viewing program from an insider's perspective.

31. *The Remote Viewers.* This report was first made public in Schnabel's book. Those of us involved in these unusual projects have discussed this incident for a long time as an example of the failure to be able to follow through on proven success.

32. Robert Jahn and Brenda Dunne, *Margins of Reality: The Role of Consciousness in the Physical World* (Harcourt, Brace, Jovanovich, 1987). This book comes as close as anyone has to providing a theory. It also contains evidence of precognitive remote viewing.

33. Commander L. R. Bremseth, "Unconventional Human Intelligence Support: Transcendent and Asymmetric Warfare Implications of Remote Viewing," Marine War College, April 28, 2001.

34. John Alexander, "Uri's Impact on the U.S. Army," 1996, posted on www.urigeller.com, contains more information about PK events.

35. Jahn and Dunne, *Margins of Reality.*

15. WINNING WORLD WAR X

1. This issue was first addressed in an op-ed piece in the *Las Vegas Sun* on October 18, 2001. This piece is included in this book as the appendix.

2. *Understanding Islam and the Muslims,* Islamic Affairs Department, Embassy of Saudi Arabia, Washington, D.C.

3. Robert D. Kaplan, *Warrior Politics: Why Leadership Demands a Pagan Ethos* (Random House, 2002).

4. Thomas L. Friedman, *The Lexus and the Olive Tree* (Farrar, Straus, and Giroux, 1999).

5. John B. Alexander, *Future War: Non-Lethal Weapons in Twenty-first-Century Warfare* (St. Martin's Press, 1999), Appendix A.

6. "The Man Who Knew," *Frontline*, PBS, October 3, 2002.

7. John B. Alexander, "A Proposal for Non-Lethal Force Development," LA-CP 94-190, the Defense Science Board, Beckman National Academy of Science Center, August 23, 1994.

8. Council on Foreign Relations, "Terrorism: Q & A," http://www.terrorismanswers.com/policy/foreignaid.html.

INDEX